A "Hands On"
approach to
teaching...

Statistics, Probability and Graphing

Linda Sue Brisby

Andy Heidcmann

Natalie Hernandez

Jeanette Lenger

Ron Long

Petti Pfau

Scott Purdy

Sharon Rodgers

HANDS ON, INC.   SOLVANG, CALIFORNIA

Layout and Graphics: Scott Purdy
Illustrators: Barbara Johnson and Suzi Matthies
Cover Art: Petti Pfau

Second Printing: December, 1990

Order Number: HO 101
ISBN 0-927726-00-9

HANDS ON, INC.
2121 Rebild Drive
Solvang, CA 93463

Introduction

This book was compiled by a group of kindergarten through eighth grade teachers at Solvang Elementary School in Solvang, California. Its purpose is to fill a void which we were experiencing at our school, and which we anticipated was being experienced at many other schools as well.

In 1985, the State of California released a document entitled, <u>Mathematics</u> <u>Framework</u> <u>For</u> <u>California</u> <u>Public</u> <u>Schools.</u> The document was revolutionary in that it sought to restructure the process of teaching math in the classroom. State commitment was so strong that all textbooks submitted for state adoption were rejected by the state textbook committee.

Classroom teachers were left in a state of confusion. Math strands, manipulatives, problem solving, cooperative learning, and calculators were "IN" (emphasized). Algorithms, memorization, pencil and paper math, and standardized tests were "OUT" (de-emphasized)! All of this occurred without the help of textbooks to bridge the transition.

At Solvang School we have been involved in a "hands-on" approach to math for a number of years. Still, we were caught in the situation of not knowing how to handle the directions of the Framework.

To fill this void, we have created this set of activity books.

This is the first in a series of seven books dealing with the strands outlined in the <u>State</u> <u>Math</u> <u>Framework.</u> It is not a textbook, but it is an invaluable supplement to your mathematics program.

All lessons described in this book are "activity based." We feel strongly that children learn best when they have concrete experiences in learning mathematical concepts.

Our approach is to provide a TASK ANALYSIS of the skills that children need to understand PROBABILITY, STATISTICS, AND GRAPHING, and to give a variety of activities which allow children to learn these skills. Activities are organized from basic to complex within each task analysis item and each lesson has a list of necessary materials, recommended classroom organization, and a basic explanation of the lesson format. We have also included extensions in many lessons.

This book was written BY TEACHERS FOR TEACHERS, and we use these activities in our classrooms every day. All activities involve the use of easily obtainable and inexpensive objects as manipulatives. There is no need to spend large sums of money to teach math. We also feel that this approach enhances the "real world" applications of our lessons. We have left out the typical flow charts, color coding, and cross reference pages that often accompany multi-grade level texts. We have included only practical, teacher based information that you can read once and use.

What we have provided is over 150 organized, concise, activity oriented lessons for teaching STATISTICS, PROBABILITY, AND GRAPHING. We have included kindergarten through eighth grade lessons so any elementary teacher will have access to remedial and enrichment lessons in one convenient source.

Statistics Probability and Graphing
Task Analysis

REAL GRAPHS	1. Uses real objects to make graphs comparing two things.	Primary
REAL GRAPHS	2. Uses real objects to make graphs comparing three or more things.	Primary
PICTURE GRAPHS	3. Makes picture graphs comparing two groups.	Primary
PICTURE GRAPHS	4. Makes picture graphs comparing three or more groups.	Primary
SUMMARIZES DATA	5. Records and summarizes data for an event.	Primary
SYMBOLIC GRAPHS	6. Makes symbolic graphs comparing groups.	Primary
CONCLUSIONS AND PREDICTIONS	7. Draws conclusions and makes predictions from data.	Primary
BAR GRAPHS	8. Colors squares of a grid to correspond to given numbers.	Primary
BAR GRAPHS	9. Interprets bar graphs as to which shows more, which show fewer.	Primary
PROBABILITY	10. Takes sampling of objects and predicts outcomes.	Primary
BAR GRAPHS	11. Completes and/or constructs bar graphs given data.	Primary/Middle
BAR GRAPHS	12. Reads and interprets bar graphs, both vertically and horizontally.	Middle
BAR GRAPHS	13. Generates an original set of data and constructs a bar graph.	Middle
PICTOGRAPHS	14. Reads and interprets pictographs.	Middle
PICTOGRAPHS	15. Constructs a pictograph given data.	Middle/Upper
PICTOGRAPHS	16. Generates data and constructs an original pictograph.	Middle/Upper
TABLES	17. Reads, interprets, and constructs tables for organizing data and solving simple problems.	Middle/Upper
AVERAGE (MEAN)	18. Manipulates real objects, bar graphs, or pictographs to find averages (mean).	Middle/Upper
AVERAGE (MEAN)	19. Computes mean or average from given data.	Middle/Upper
RANGE/MEDIAN/MODE	20. Identifies range, median, and mode from given data.	Middle/Upper

PROBABILITY	21. Writes fractions to represent probability.	Middle/Upper
DOUBLE BAR GRAPHS	22. Reads, constructs, and interprets double bar graphs.	Middle/Upper
BROKEN LINE GRAPHS	23. Reads and interprets broken line graphs.	Middle/Upper
BROKEN LINE GRAPHS	24. Constructs a broken line graph.	Middle/Upper
CIRCLE GRAPHS	25. Reads and interprets circle graphs.	Upper
TREE DIAGRAMS	26. Uses a tree diagram to find total number of possible outcomes.	Upper
TREE DIAGRAMS	27. Draws a tree diagram for a given problem.	Upper
PROBABILITY	28. Predicts results of simple probability experiments by multiplying the probability of favorable outcomes by the total number of outcomes.	Upper
PROBABILITY	29. Identifies zero probability as an event which cannot occur and a probability of one as an event which is certain to take place.	Upper
SAMPLES/CENSUS	30. Identifies whether to use a sample or a census in a given situation.	Upper
CIRCLE GRAPHS	31. Constructs a circle graph given data.	Upper
LINE OF BEST FIT	32. Makes predictions based on line of best fit.	Upper
TABLES	33. Completes and constructs frequency tables.	Upper
PROBABILITY	34. Uses multiplications to find total number of outcomes.	Upper
PROBABILITY	35. Calculates the probability when choices are dependent or independent.	Upper
SCATTERGRAMS	36 Reads and makes a scattergram and tells whether it shows positive, negative, or no correlation.	Upper
CURVED GRAPHS	37. Reads, interprets, and constructs curved graphs.	Upper
GRAPHS	38. Identifies and constructs appropriate graphs for a given situation.	Upper

Table of Contents

Do You Like School?
September
Grade Level: Primary

TASK ANALYSIS: 6 – Makes symbolic graphs comparing groups
7 – Draws conclusions and makes predictions from data

MATERIALS: Marking pens, masking tape, YES/NO words out of butcher paper for graph.

ORGANIZATION: Whole class activity
Kindergarten/primary: 20 to 30 minutes

PROCEDURE: – Ask children the question, "Do you like school?"
– Using the YES/NO graph have them write their names inside of the yes or no.
– Count the number of names in each.
– Ask questions such as, "Which one has the most names?" "Which one has the least names?" "How many more 'yes' answers are there than 'no' answers?"
– "Do you think the results would be the same if we graphed this question on another day?"

| Graphing |
| Comparing |
| Counting |
| Interpreting |
| Recording |
| Predicting |

What Kind of Apple Did You Bring?

September – Apple Math
Grade Level: Primary

TASK ANALYSIS: 1 – Uses real objects to make graphs comparing two things
3 – Makes picture graphs comparing two groups
5 – Records and summarizes data for an event
7 – Draws conclusions and makes predictions from data
8 – Colors squares of a grid to correspond to given numbers
9 – Interprets bar graphs as to which shows more and fewer

MATERIALS: Red or green apples brought by students, graphing plastic for floor graphs, tents with a picture of a red and green apple, butcher paper for picture graph, one xeroxed apple for each child, red and green crayons, one xeroxed bar graph page for each child, masking tape, glue stick.

ORGANIZATION: Whole class activity
Kindergarten: 1 to 2 days; Primary: 50 to 60 minutes

PROCEDURE: – Make tents for real graph (graphing plastic).
– Make butcher paper picture graph and divide in half with headings.

> Graphing
> Comparing
> Counting
> Interpreting
> Recording

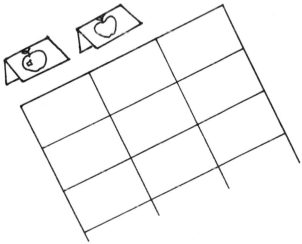

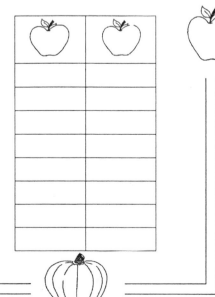

- Have each child get "to know" his apple -- color, size, texture, etc.
- Have children place red apples on floor graph under proper heading.
- Have children with green apples do the same.
- Count the apples in each category.
- Compare: which has more, less, how many more ___ apples than ___ apples?
- Have each child remove his apple from the graph.

- Give each child a "xeroxed apple" and have him color it the same color as his real apple.
- Have each child bring his apple and glue it to the picture graph.
- Count and compare as above.
- Ask if both graphs were the same or different.

- Give each child a xeroxed page for his own bar graph.
- Have children copy data from picture graph.
- Children conclude and predict from data.

Red	Green

Can You Estimate the Circumference of Your Apple?

September – Apple Math
Grade Level: Primary

TASK ANALYSIS: 2 – Uses real objects in graphs comparing three or more things
4 – Makes picture graphs comparing three or more groups
5 – Records and summarizes data for an event

MATERIALS: Apples brought by children, string wrapped around tongue depressors, scissors, masking tape, headings for graphs.

ORGANIZATION: In kindergarten with 2 or 3 children at a station, 20 to 30 minute activity. In primary: whole class, 20 to 30 minute activity.

PROCEDURE: – Make headings.
– Put masking tape under headings (sticky side out).

Graphing
Comparing
Counting
Measuring
Recording
Estimating

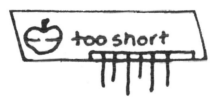

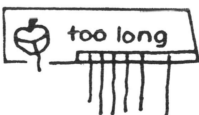

– Pass string-wrapped tongue depressor to each child and have him estimate the length of string it will take to go around his apple.
– Cut the string that was estimated.

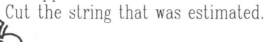

– Measure the apple with the string.
– Put the string under the correct heading.
– Count the number of strings under each heading.
– Interpret data collected.

– For kindergarten, while entire group works on a class activity, take two or three children at a time to measure their apples.
– After all have measured and placed string under proper heading, bring whole class together to count and interpret data.

EXTENSIONS: Weigh the apples
                   Estimate seeds
                   Make applesauce or apple pie
                   Read Johnny Appleseed
                   Describe your apple (a writing lesson)
                   Art projects: Apple Print Pictures
                   Apple Tree Seasons

 # Apple Art

The Apple Tree Cycle

1. Glue 4 pieces of brown paper on your paper to make tree trunks.

2. Use a brown crayon and draw branches on the first tree for <u>winter</u>.

3. Sponge pink paint on the second tree for <u>spring</u>.

4. Sponge green paint on the third tree and print red apples with a piece of carrot for <u>summer</u>.

5. Sponge red, yellow and orange paint on the last tree for <u>fall</u>.

Make a large apple-shaped chart with these instructions ↗

You will need:

- 12" x 18" construction paper
- 4" x 1" brown construction for tree trunks (4 per child)
- brown crayons
- pink, green, red, orange & yellow tempera paint
- carrot slices
- sponge pieces for each color of paint

What Kind of Fish Are In The Bag?

October
Grade Level: Primary

TASK ANALYSIS: 10 – Takes sampling of objects and predicts outcome

MATERIALS: Bag of pretzel fish, bag of cheddar fish, 1 lunch bag, butcher paper, marking pen, glue

ORGANIZATION: Whole class activity
Kindergarten/primary: 45 to 50 minutes

PROCEDURE: – Put five pretzel fish and one cheddar fish into paper bag (without children seeing them).
– Have each child in class cover his eyes and draw out one fish at a time while another child tallies the number of draws on butcher paper.
– After a child chooses, the fish is returned to the bag.

 Ⅲ茶 茶Ⅱ 茶Ⅱ 茶Ⅱ 茶Ⅱ

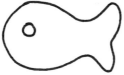 茶Ⅱ Ⅰ

– Record child's choice by gluing same type of fish on butcher paper graph (keep an extra bag of pretzel and cheddar chips beside graph).
– Tell the children there are six fish in the bag.
– By looking at the information on the graph, ask, "How many fish in the bag will be pretzel and how many will be cheddar?"
– Read and interpret.

EXTENSIONS: – Graph may be real, pictorial, or tally marks on chalkboard to make a symbolic graph.
              – Other objects such as macaroni, beans, beads, or unifix cubes may also be used.

How Many Vertical Lines Are on the Pumpkin?

October – Pumpkin Math
Grade Level: Primary

TASK ANALYSIS: 7 – Draws conclusions and makes predictions from data
8 – Colors squares of a grid to correspond to given numbers

MATERIALS: Pumpkin, watercolor pen, colored paper squares, crayons or pencils, glue stick, column graph.

ORGANIZATION: Whole class activity
Kindergarten/primary: 20 to 30 minutes

PROCEDURE: – Make column graph with number headings.
– Cut colored paper squares.

Graphing
Comparing
Counting
Interpreting
Recording

6	7	8	9	10	11	12	13	14+

– Have children estimate how many vertical lines are on the pumpkin (demonstrate vertical).
– Have primary children write their estimates on paper square –– kindergarten children leave blank.

- Place the graphing plastic on the floor with the pumpkins at the top of the three columns.
- Call each group one at a time to stand on the grid.
- Count and interpret -- "Which has more?" "less?" "There are _____ more sad pumpkins than happy ones."
- Continue asking questions of this type.

- Give each child xeroxed pumpkin.
- Have children draw a face on the pumpkin as to how they want their pumpkin carved.
- Have each child bring his pumpkin and glue to graph.
- Interpret graph.
- Ask questions as above.
- "Was this graph the same or different than our real graph?"

EXTENSIONS: Pumpkin attribute game (20 questions –yes/no answers)
Weigh the pumpkins
Estimate seeds
Show place value
Estimate circumference
Art printing using seeds
Bake the seeds
Make pumpkin pie
Visit pumpkin patch

17

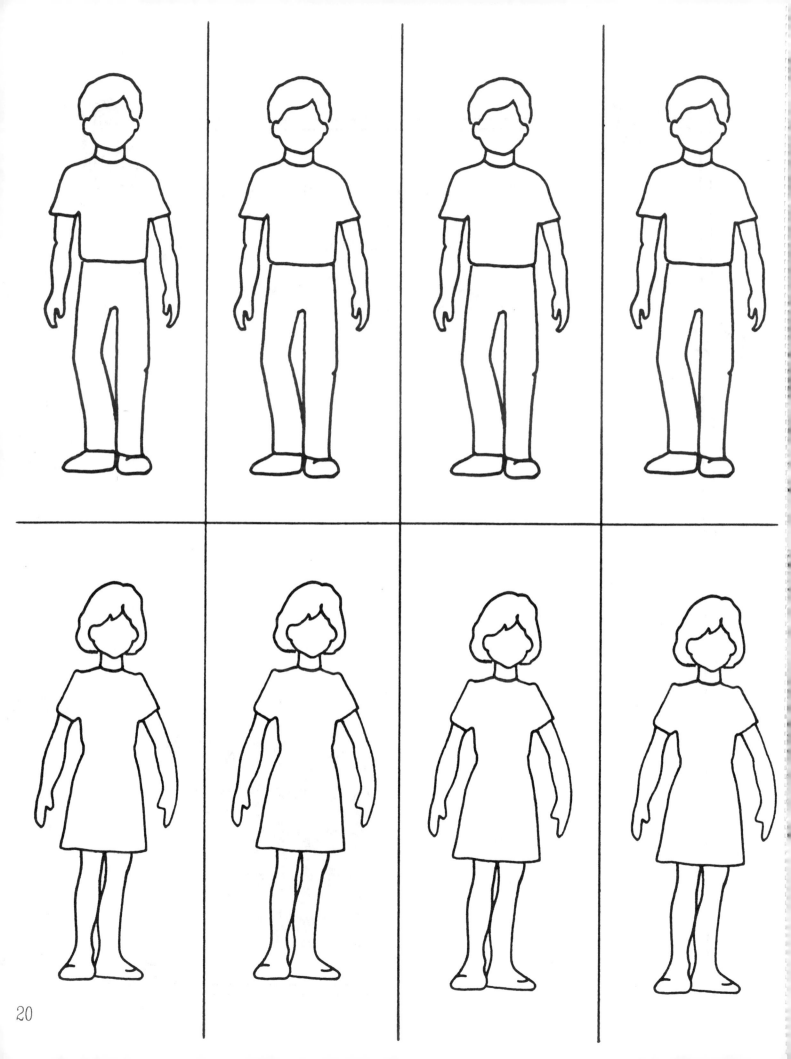

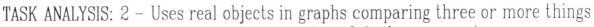

How Do You Like Your Popcorn?

November – Popcorn Math
Grade Level: Primary

TASK ANALYSIS: 2 – Uses real objects in graphs comparing three or more things
5 – Records and summarizes data for an event
6 – Makes symbolic graphs comparing groups
7 – Draws conclusions and makes predictions from data

MATERIALS: Popped corn (plain – buttered – butter and salt), cups with children's names on them, chart paper for graphs, heading tents for types of popcorn, graphing plastic, three bowls.

ORGANIZATION: Whole class activity
Kindergarten/primary: 20 to 40 minutes

PROCEDURE:
- Prepare graphs.
- Put masking tape names on plastic cups (one per child).
- Label tents for plain, buttered, butter/salt.
- Put popped popcorn in correct bowl.
- Put graphing plastic on floor; place headings at top.
  Give each child a plastic cup.
- Have children sample each type of popcorn. After each sampling, discuss the taste.

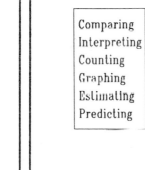

| Comparing |
| Interpreting |
| Counting |
| Graphing |
| Estimating |
| Predicting |

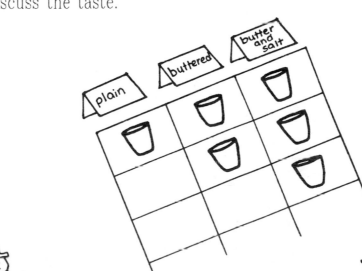

– Ask question, "Which one did you like the best?"
– Have child put cup under heading of the way he likes popcorn.
– Read and discuss graph.
– Have the children get their cups, take off label, and place label under the correct heading on the symbolic graph.
– Read and interpret.
– Have child fill cup with favorite popcorn and enjoy.

EXTENSIONS

Popcorn attribute game (20 questions – yes/no answers)
Estimate spoonfuls
Tally spoonfuls
Estimate kernels in small jar
Show "place value"
Describe popcorn – writing lesson
Read about the history of popcorn
Trace journey of popcorn plant to popcorn bowl
Discuss different ways to use popcorn
Read Popcorn Dragon
Art Project: Make a popcorn dragon
PE: Pretend you're a popcorn kernel
Make a four column graph by adding a heading of "salt only."

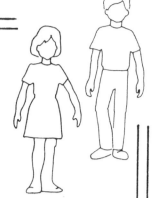

How Many People in Your Family?
November
Grade Level: Primary

TASK ANALYSIS: 6 – Makes symbolic graphs comparing groups
9 – Interprets bar graphs as to which shows more and fewer

MATERIALS: Butcher paper graph, rectangle paper with child's name on it, glue stick.

ORGANIZATION: Whole class activity
Kindergarten/primary: 20 to 30 minutes

PROCEDURE: _ Make butcher paper graph with number heading columns.
– Cut rectangles and put children's names on them.

Comparing
Interpreting
Recording
Graphing

How Many People in Your Family?

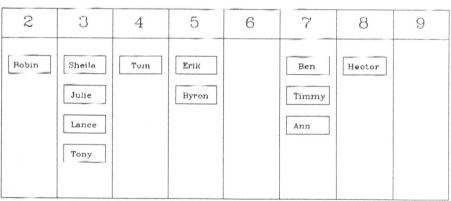

2	3	4	5	6	7	8	9
Robin	Sheila	Tom	Erik		Ben	Hector	
	Julie		Byron		Timmy		
	Lance				Ann		
	Tony						

– Discuss families and members of families.
– Pass out rectangles (with names) to children.
– Have one at a time come and glue name in correct column.
_ Interpret and discuss graph.

Do You Break a Pinata at Christmas?

December
Grade Level: Primary

TASK ANALYSIS: 1 – Uses real objects to make graphs comparing two things

MATERIALS: Wrapped hard candy (one per child), yes/no floor graph of butcher paper.

ORGANIZATION: Whole class activity
Kindergarten/primary: 20 to 30 minutes

PROCEDURE: – Make yes/no graph of butcher paper.

Comparing
Interpreting
Counting
Graphing

Have You Broken a Pinata?

YES	NO
🍬	🍬
🍬	🍬
🍬	🍬
	🍬
	🍬
	🍬
	🍬

– Explain a pinata.
– Ask the question, "Do you break a pinata at Christmas?"
– Have children put a candy on the graph under yes or no.
– Read and interpret graph by questioning.
– Have the children break a pinata.

Do You Celebrate Hanukkah or Christmas?

December
Grade Level: Primary

TASK ANALYSIS: 6 – Makes symbolic graphs comparing groups
7 – Draws conclusions and makes predictions from data

MATERIALS: Butcher paper, colored strips of construction paper – 1 1/2" by 6" (3 cm x 15cm), glue sticks.

ORGANIZATION: Whole class activity
Kindergarten/primary: 20 to 30 minutes

PROCEDURE: – Make butcher paper graph divided in half with headings.
– Cut strips of construction paper for chain links.

Comparing
Interpreting
Counting
Graphing

CHRISTMAS OR HANUKKAH

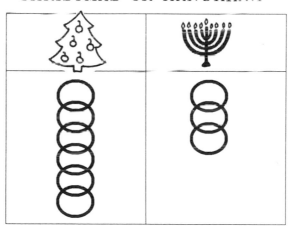

– Discuss the traditions of Hanukkah and Christmas.
– Ask question, "Do you celebrate Hanukkah or Christmas?"
– Children take a construction paper strip and make a chain link under correct heading.
– Count and interpret data by questioning.

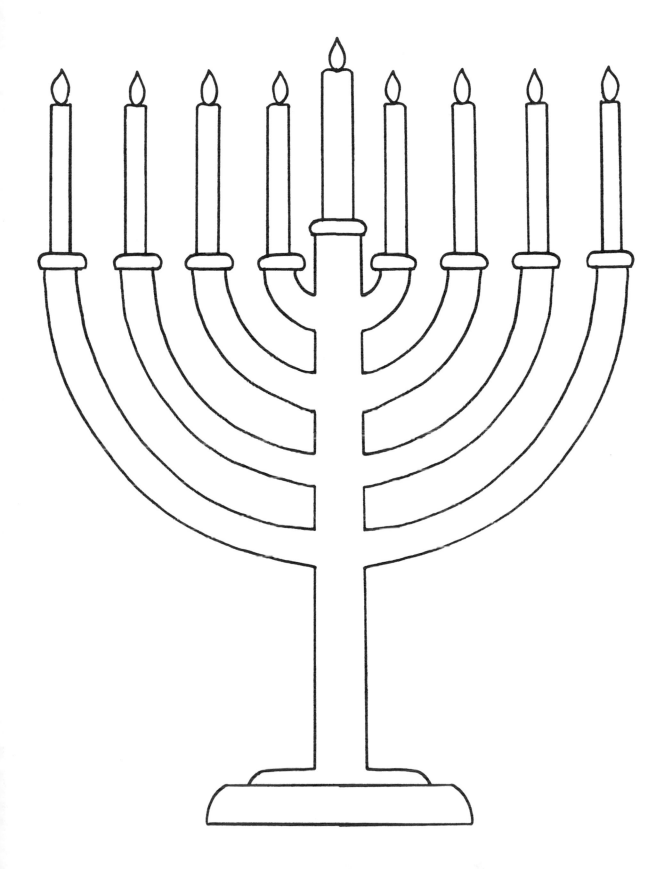

29

Which Is Your Favorite Song?

December
Grade Level: Primary

TASK ANALYSIS: 6 – Makes symbolic graphs comparing groups

MATERIALS: Butcher paper, red and green unifix cubes

ORGANIZATION: Whole class activity
Kindergarten/primary: 20 to 30 minutes

PROCEDURE: – Make graph of butcher paper with heading of "Rudolph the Red-Nosed Reindeer" and "Jingle Bells."
– Prepare tub of red and green unifix cubes.

Graphing
Comparing
Counting
Interpreting
Recording

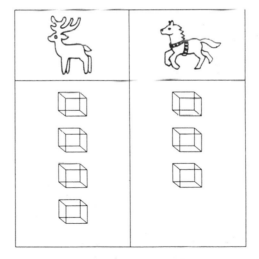

– Sing "Rudolph" and "Jingle Bells."
– Have children pick a red unifix cube if they like "Rudolph" and a green unifix cube if they like "Jingle Bells."
– Put same color unifix cubes together under correct heading.
– Read and interpret data collected.

Have You Ever Built a Snowman?

January
Grade Level: Primary

TASK ANALYSIS: 1 – Uses real objects to make graphs comparing two things
9 – Interprets bar graphs as to which shows more and fewer
11 – Completes and constructs bar graphs given data

MATERIALS: Marshmallows, toothpicks, graphing plastic, "YES/NO" heading tents, crayons, xeroxed bar graphs (one per child).

ORGANIZATION: Whole class activity
Kindergarten/primary: 20 to 30 minutes

PROCEDURE: – Make headings: yes/no.
– Xerox bar graphs (one per child).

Comparing
Interpreting
Counting
Graphing

I've Made a Snowman!

YES	NO

– Ask children, "Have you ever built a snowman?"
– Children make marshmallow snowmen.
– Gather at floor graph.
– Children place snowmen on graphing plastic under proper heading.
– Read and interpret.
– Have children record results of the floor graph on individual bar graphs.
– Read and interpret.

Have you ever built a snowman?

Yes	No

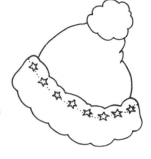

What Color Hat Does Your Snowman Wear?

January
Grade Level: Primary

TASK ANALYSIS: 4 – Makes picture graphs comparing three or more groups
5 – Records and summarizes data for an event
6 – Makes symbolic graphs comparing groups
7 – Draws conclusions and makes predictions from data
11 – Completes and constructs bar graphs given data

MATERIALS: Butcher paper, hat page (two hats per child), snowman page (one snowman per child), crayons, glue stick, scissors, grid page (one per child).

ORGANIZATION: Whole class activity
Kindergarten/primary: 20 to 30 minutes

PROCEDURE: – Make butcher paper graph with six columns and hat headings.
– Xerox hat, snowman and grid pages.

Comparing
Interpreting
Graphing
Recording
Predicting

WHAT COLOR HAT FOR YOUR SNOWMAN?

RED	BLUE	GREEN	YELLOW	BLACK	ORANGE

– Discuss the six colors to be used.
– Ask question, "What color hat do you want your snowman to wear?"
– Give each child one hat.
– Have children color, cut and glue hats on picture graph.
– Read and interpret graph.
  Have children record data from picture graph on bar graph (grid page).
– Read and interpret.
– Have children color remaining hat to put on snowman picture.

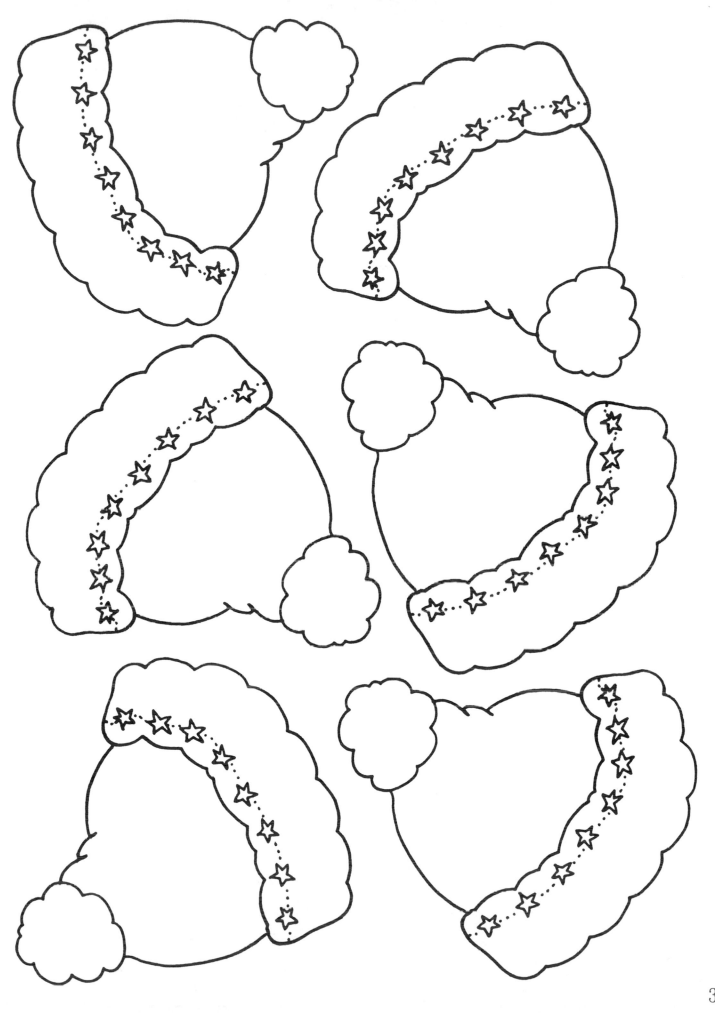

Have You Ever Been to the Snow?

January
Grade Level: Primary

TASK ANALYSIS: 3 – Makes picture graphs comparing two groups
5 – Records and summarizes data for an event

MATERIALS: Blackline master of a snowflake (one copy per child), butcher paper, glue stick, scissors, yes/no heading.

ORGANIZATION: Whole class activity
Kindergarten/primary: 20 to 30 minutes

PROCEDURE: – Duplicate snowflake (one per child).
– Make butcher paper graph divided in half with yes/no headings.

Comparing
Interpreting
Recording
Graphing

Have You Ever Been to the Snow?

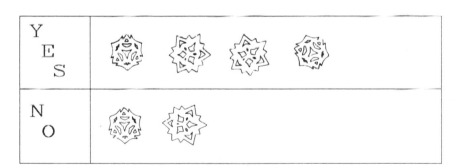

Y E S	🔲	🔲	🔲	🔲
N O	🔲	🔲		

– Discuss the attributes of snow.
– Have children cut out snowflakes.
– Glue snowflakes to graph beside proper heading.
– Read and interpret.

Snowflakes

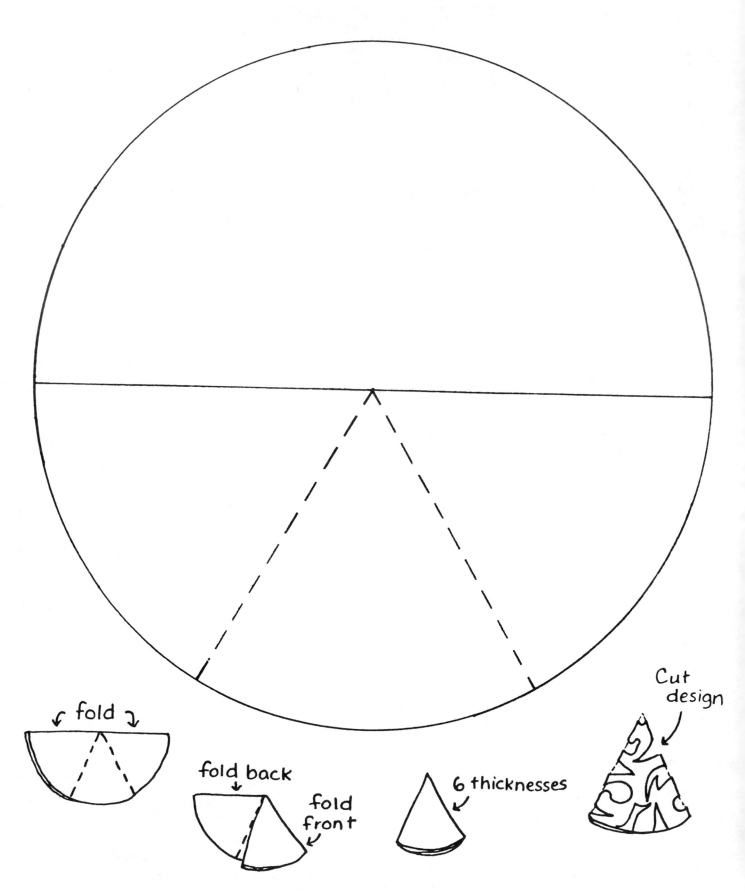

fold

fold back

fold front

6 thicknesses

Cut design

How Many Ways Can You Spill The Beans?

January
Grade Level: Primary

TASK ANALYSIS: 10 – Takes sampling of objects and predicts outcome

MATERIALS: Large lima beans (4 per child), spray paint, styrofoam cups (1 per child), spray paint, crayons, bean sheet (1 per child or small group)

ORGANIZATION: Pairs or small groups
Kindergarten (with guidance)/primary: 30 to 45 minutes

PROCEDURE:
- Spray paint beans on one side only.
- Ask the question, "If you spill the beans and record how many painted sides come up each time, do you think you will get one color more than the other?" "If so, which color will it be?" "Why do you think it will be that color?"
- Have children predict which side will come up most often.
- Have children spill the beans at least eight times.
  Have children record actual results on bean sheet.
- Ask, "Was your prediction accurate?"

EXTENSIONS: Try it with other numbers of beans.

Would Abe Lincoln Look Better With or Without a Beard?

February
Grade Level: Primary

TASK ANALYSIS: 3 – Makes picture graphs comparing two groups
5 – Records and summarizes data for an event

MATERIALS: Butcher paper, headings (with a beard, without a beard), drawing paper squares 4" by 4" (10 cm x 10 cm), crayons, glue stick, a picture of Abraham Lincoln.

ORGANIZATION: Whole class activity
Kindergarten/primary: 20 to 30 minutes

PROCEDURE: – Make butcher paper graph divided in half with headings.
– Cut paper squares.

Comparing
Interpreting
Recording
Graphing

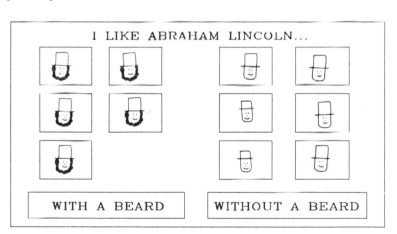

I LIKE ABRAHAM LINCOLN...

WITH A BEARD   WITHOUT A BEARD

– Discuss Abe.
– Explain and model "U-shaped" face for Abe.
– Ask question, "Would Abe look better with or without a beard?"
– Have children draw Abe on four inch squares.
– Have children glue drawing above the correct heading.
– Read and interpret graph.

What Color Are Your Hearts?

February
Grade Level: Primary

TASK ANALYSIS: 2 – Uses real objects in graphs comparing three or more things
4 – Makes picture graphs comparing three or more groups
5 – Records and summarizes data for an event
6 – Makes symbolic graphs comparing groups
7 – Draws conclusions and makes predictions from data
8 – Colors squares of a grid to correspond to given numbers
9 – Interprets bar graphs as to which shows more and fewer
11 – Completes and constructs bar graphs given data

MATERIALS: Large candy conversation hearts (three or more bags), cups for each group, pencils, crayons, prediction graph for whole group, counting graph for each pair of children, one xeroxed heart per child, one "Valentine Hearts Graph" per pair, and sorting sheets.

ORGANIZATION: Whole class activity
Kindergarten/primary: 2 day activity; 20 to 30 minutes per day

PROCEDURE: – Prepare prediction graph and counting graph.
– Have children predict which color heart will appear most often in the bags of candy.
– Have them write their names on a paper heart and glue under proper heading on the prediction graph.
– Read and interpret.

Comparing
Interpreting
Recording
Graphing

Paper hearts cut out
by children

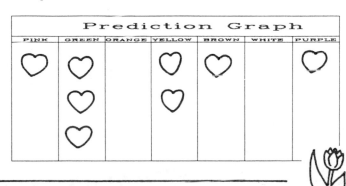

Prediction Graph

PINK	GREEN	ORANGE	YELLOW	BROWN	WHITE	PURPLE

45

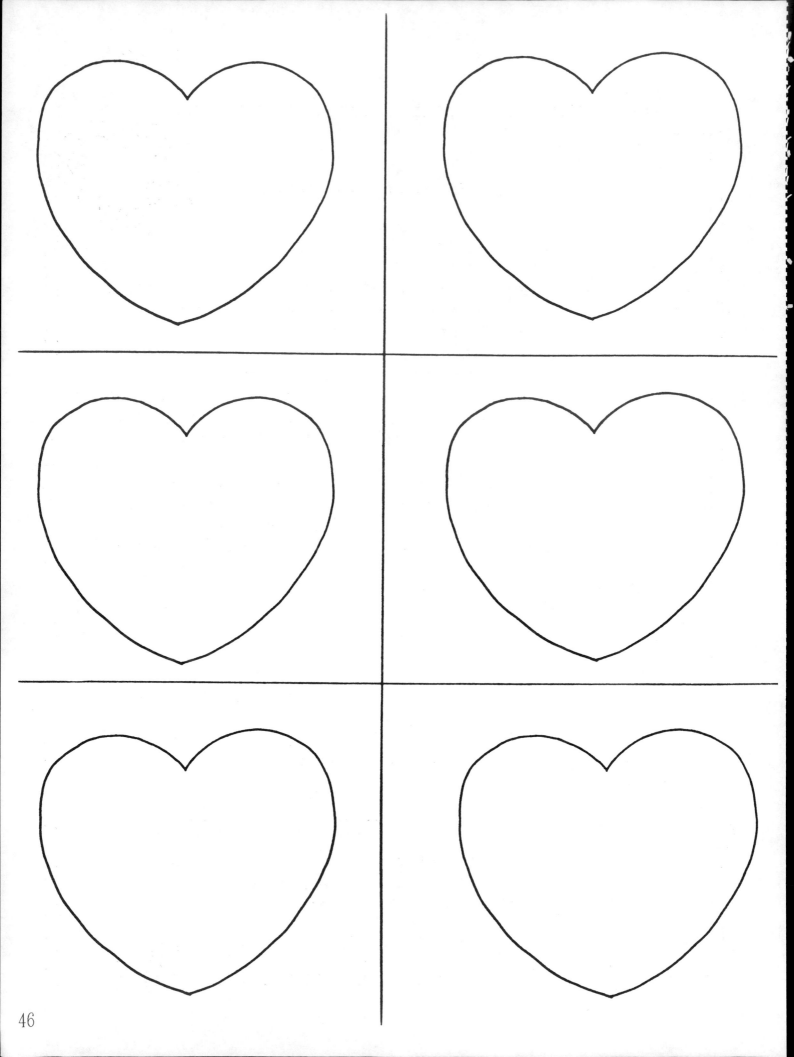

46

- Divide the class into pairs and give each pair a cup full of hearts.
- Children sort and classify using sorting sheets.
- Each pair records data with correct color crayon on his bar graph.
- Record complete data from each pair on class counting graph, periodically summarizing student observations ("Which is less?" "There are more ____ than ____ .").

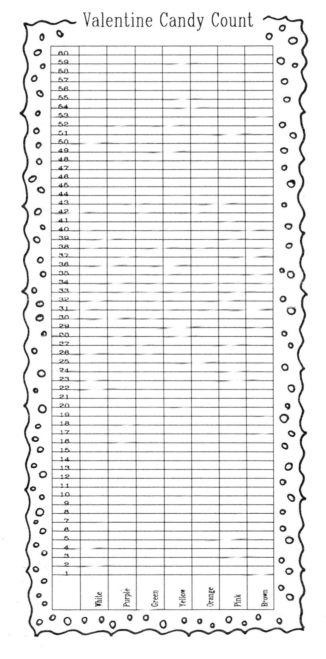

Valentine Candy Count

47

Valentine Hearts Graph

Number of Hearts	White	Purple	Green	Yellow	Orange	Pink	Brown

Which February Day Did You like the Best?

February
Grade Level: Primary

TASK ANALYSIS: 6 – Makes symbolic graphs comparing groups
7 – Draws conclusions and makes predictions from data
8 – Colors squares of a grid to correspond to given numbers
9 – Interprets bar graphs as to which shows more and fewer

MATERIALS: Column graph with picture headings, red construction paper squares (3" by 3") with children's names on them, crayons, glue stick, butcher paper graph to correspond with column graph

ORGANIZATION: Whole class activity
Kindergarten/primary: 20 to 30 minutes

PROCEDURE: – Prepare graphs.
– Cut squares of red construction paper and put children's names on them.
– Discuss each special day: Susan B. Anthony Day, Valentine's Day, Lincoln's Birthday, Washington's Birthday, the 100th day of School, and the 29th (leap year).
– Ask the question, "What was your favorite day?"

Graphing
Comparing
Counting
Interpreting
Recording
Predicting

What is Your Favorite Day in February ?

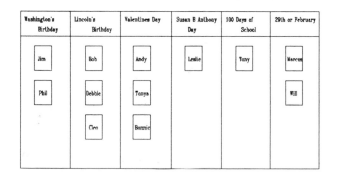

Washington's Birthday	Lincoln's Birthday	Valentines Day	Susan B Anthony Day	100 Days of School	29th or February
Jim	Bob	Andy	Leslie	Tony	Marcus
Phil	Debbie	Tonya			Will
	Cleo	Bonnie			

Favorite Day Bar Graph

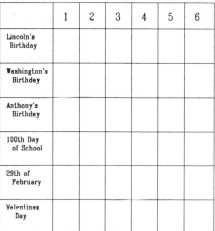

	1	2	3	4	5	6
Lincoln's Birthday						
Washington's Birthday						
Anthony's Birthday						
100th Day of School						
29th of February						
Valentines Day						

- Give children their paper squares.
- Have children glue onto column graph.
- Count the number of papers in each column.
- Read and interpret.
- Take data from column graph and record to bar graph.
- Interpret bar graph as to which shows more, less, fewer, and same.

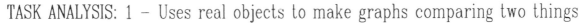

Which Kite Would You Like to Fly?

March
Grade Level: Primary

TASK ANALYSIS:
1 – Uses real objects to make graphs comparing two things
2 – Uses real objects in graphs comparing three or more things
3 – Makes picture graphs comparing two groups
4 – Makes picture graphs comparing three or more groups
5 – Records and summarizes data for an event

MATERIALS: Two or three real kites (box kite, diamond kite, dragon kite), unifix cubes, xerox kite page – one per child.

ORGANIZATION: Whole class activity
Kindergarten/primary: 20 to 30 minutes

PROCEDURE: – Discuss attributes of each kite.
– Ask the question, "Which kite would you like to fly?"
– Have children stack unifix cubes under "real kite" representing their choice.
– Give each child a copy of the kite picture page.
– Have child reproduce "real" bar graph as a picture graph.

> Comparing
> Interpreting
> Recording
> Graphing

OUR FAVORITE TYPES OF KITES

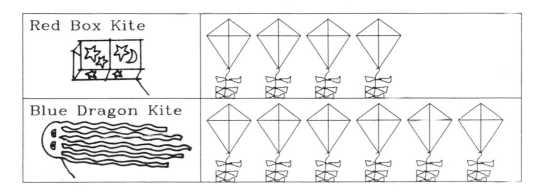

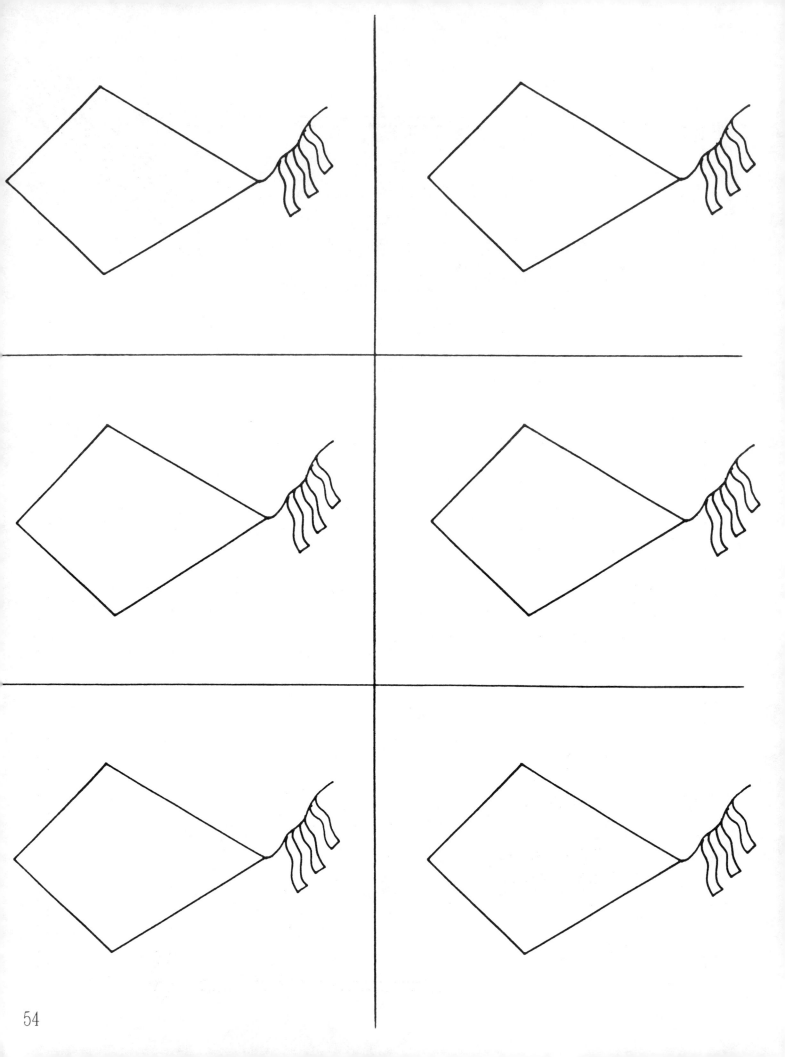

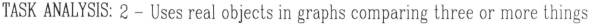

How Many Peanuts in Your Shell?

March
Grade Level: Primary

TASK ANALYSIS: 2 – Uses real objects in graphs comparing three or more things
5 – Records and summarizes data for an event
10 – Takes sampling of objects and predicts outcome

MATERIALS: Butcher paper wall graph, bag of peanuts in the shell, a bowl of peanuts for recording, glue, tents labeled 0 – 1 – 2 and 3 or more, two counting cups per child.

ORGANIZATION: Whole class activity
Kindergarten/primary: 20 to 30 minutes

PROCEDURE: – Make column graph of butcher paper with headings.

Comparing
Interpreting
Recording
Graphing

HOW MANY PEANUTS IN THE SHELL?

Peanuts glued on
by children

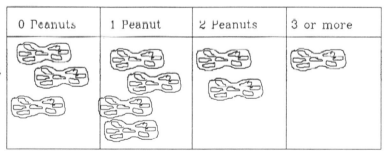

0 Peanuts	1 Peanut	2 Peanuts	3 or more

– Introduce bag of peanuts by playing, "What's in my bag?" (Yes/no answers).
– Ask students the probability questions, "If we each opened a peanut shell, how many peanuts do you predict we would find most often?"
– Each child takes a peanut from the bowl and glues to graph in the proper column.

– Read and interpret data.
– Put tent headings on floor.
– Have children take one peanut shell, open, and put peanuts found into counting cup.
– Have each child put counting cup under correct tent heading.
– Read and interpret.
– Ask questions, "Will our graph remain the same or will it change if we each open another peanut?" Discuss responses.
– Repeat above procedure of children opening and placing peanuts.
– Read and interpret.
– Ask, "How close was your prediction?"

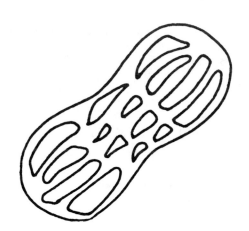

How Do You Like Your Peanuts?

March
Grade Level: Primary

TASK ANALYSIS: 2 – Uses real objects in graphs comparing three or more things
5 – Records and summarizes data for an event
8 – Colors squares of a grid to correspond to given numbers
9 – Interprets bar graphs as to which shows more and fewer

MATERIALS: A jar of peanut butter, a can of peanuts, roasted peanuts in the shell, xerox of one grid per child, counting cups for each child, graphing plastic, crayons, knife.

ORGANIZATION: Whole class activity
Kindergarten/primary: 20 to 30 minutes

PROCEDURE:
– Give each child a sample of his favorite peanut in counting cup.
– Have child place under correct heading on graphing plastic.
– Read and interpret.
– Give each child a peanut grid sheet.
– Have children record data from graphing plastic onto grid sheet using a crayon.
– Read and interpret.

Comparing
Interpreting
Recording
Graphing

Name _____

Peanuts!	Peanut Butter!	In the Shell

How Can You Get Your Eggs To School Without Cracking?

April
Grade Level: Primary

TASK ANALYSIS: 2 – Uses real objects in graphs comparing three or more things
5 – Records and summarizes data for an event
7 – Draw conclusions and makes predictions from data

MATERIALS: Graphing plastic, headings with real items (egg cartons, popcorn, styrofoam, etc.), hard-boiled eggs brought by children, tents (cracked, not cracked).

ORGANIZATION: Whole class activity
Kindergarten/primary: 2 day activity; 30 minutes per day

PROCEDURE: – Make headings for the types of packing materials.
– Make tents (cracked, not cracked).

Comparing
Interpreting
Recording
Graphing

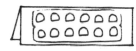

– Brainstorm on the first day, different ways in which to safely bring hard-boiled eggs to school (be creative).

– Generate categories on the second day and make headings (be prepared per discussion of previous day).
– Have children place packed eggs in containers under appropriate headings on graphing plastic.
– Read and interpret.

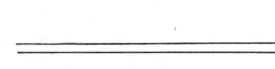

- Have children get their eggs from graphing plastic.
- Place new headings (cracked/not cracked) on graphing plastic.
- Have children place their eggs under proper heading.
- Read and interpret.

What Color Did You Color Your Egg?

April
Grade Level: Primary

TASK ANALYSIS:
2 – Uses real objects in graphs comparing three or more things
5 – Records and summarizes data for an event
7 – Draws conclusions and makes predictions from data
8 – Colors squares of a grid to correspond to given numbers
9 – Interprets bar graphs as to which shows more and fewer

MATERIALS:
Hard boiled eggs brought by children, egg cartons, tissue paper cut into small pieces, starch, paint brushes, tent headings (blue/yellow, green/red, speckled), xeroxed grids (one per child).

ORGANIZATION:
Whole class activity
Kindergarten/primary: 2 day activity; 30 to 40 minutes per day.

PROCEDURE:
- Prepare egg cartons with tent headings.
- Cut tissue paper into small pieces.
- Xerox grid page one per child.

<table>
<tr><td>Comparing
Interpreting
Recording
Graphing</td></tr>
</table>

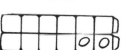 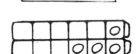

- Paint eggs with starch and apply color tissue paper then peel off tissue paper and let egg dry overnight.

- On the second day, ask "What color did you color your egg?"
- Have children put egg in proper carton under correct heading.
- Read and interpret.

- Give each child xeroxed grid sheet and have him color grid to match egg carton graph.
- Read and interpret.

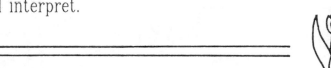

How did you color your eggs?

blue/ yellow		green / red		speckled	

What is Your Favorite Color Tulip?
April
Grade Level: Primary

TASK ANALYSIS: 4 – Makes picture graphs comparing three or more groups
5 – Records and summarizes data for an event
6 – Makes symbolic graphs comparing groups
7 – Draws conclusions and makes predictions from data

MATERIALS: Xerox one tulip and stem per child, make wooden shoes of tagboard of as many colors as you wish to graph, glue stick, crayons, scissors, unifix cubes, grid page (one per child).

ORGANIZATION: Whole class activity
Kindergarten/primary: 20 to 40 minutes

PROCEDURE: – Make tagboard shoes (enlarge pattern).

| Comparing |
| Interpreting |
| Recording |
| Graphing |

– Brainstorm about Holland bringing out the idea of tulips and wooden shoes.
– Have children color tulip their favorite color.
– Have children color stem and leaves green.
– Glue tulip to stem.
– Have children glue flower in proper wooden shoe.
– Read and interpret.

EXTENSION: – Have children record data from wooden shoes with unifix cubes.
– Read and interpret.
– Have children take unifix data and record to grid with crayons.
– Read and interpret.

What is Your Favorite Color Tulip?

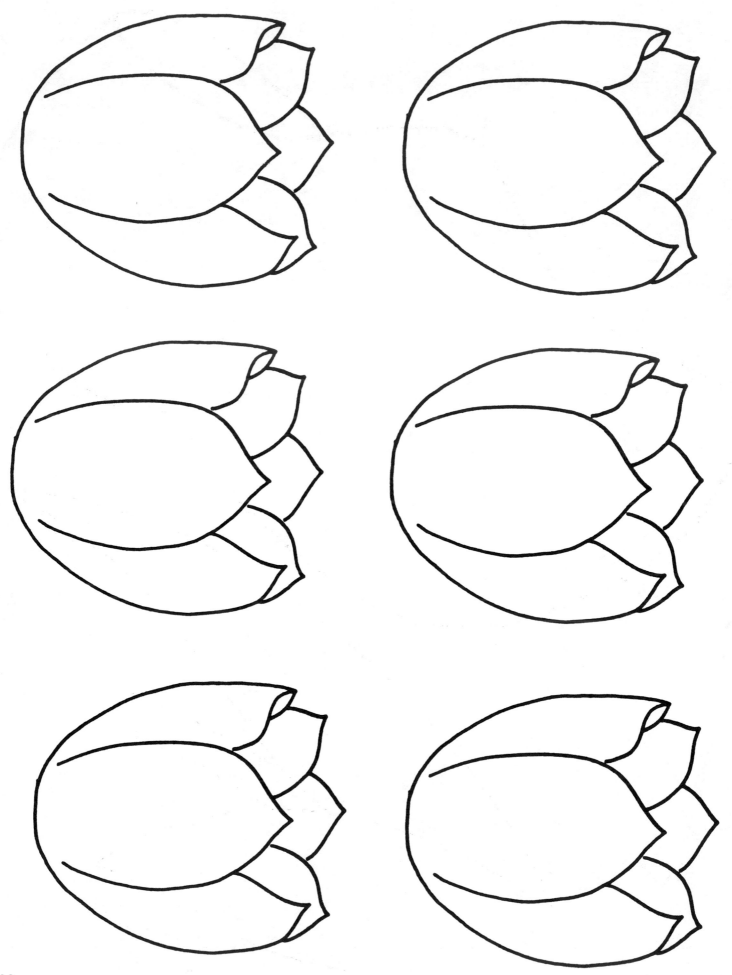

What Is Your Favorite Sign of Spring?

April
Grade Level: Primary

TASK ANALYSIS: 4 – Makes picture graphs comparing three or more groups
5 – Records and summarizes data for an event

MATERIALS: Butcher paper, 5" by 5" squares of white drawing paper (one per child), black marker to write headings, glue stick.

ORGANIZATION: Whole class activity
Kindergarten/primary: 60 minutes

PROCEDURE: – Cut squares.
– Prepare butcher paper graph.

Comparing
Interpreting
Recording
Graphing

SIGNS OF SPRING

Animals	Flowers	Trees	Weather	Other

– Go on spring walk.
– Ask questions such as, "Does it smell any different now?" "Do you see or hear animals?" " Is the weather different?" "How is this tree different now than in fall or winter?"
– Brainstorm for signs of spring after spring walk.
– Have children draw favorite sign of spring on white square.
– Glue squares under proper heading on graph.
– Read and interpret.

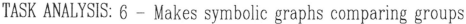

How Does Your Plant Grow?

May – Plant Math
Grade Level: Primary

TASK ANALYSIS: 6 – Makes symbolic graphs comparing groups
7 – Draws conclusions and makes predictions from data
8 – Colors squares of a grid to correspond to given numbers
9 – Interprets bar graphs as to which shows more and fewer
11 – Completes and/or constructs bar graphs given data
12 – Reads and interprets bar graphs both vertically and horizontally

MATERIALS: School milk carton or waxed cup for each child, potting soil. lima bean seeds, one to ten template, 12″ by 18″ white paper (one per child), crayons, glue, construction paper strips 1/2″ by 9″

ORGANIZATION: Whole class activty
Kindergarten/primary: two weeks 20 minutes per day

PROCEDURE: – Prepare cups for planting (cut tops off milk cartons).
– Plant seeds and have children prepare individual graphs and record growth each day.
– Read and interpret daily.

Graphing
Comparing
Counting
Interpreting
Recording
Predicting

A Suggestion: Plant seeds on a Wednesday or Thursday to give consecutive days to observe growth.

Day 4	Seth			
Day 5	Ben	Larry		
Day 6	Robin	Gene	Sven	
Day 7	Byron	Sara	Emily	Rex
Day 8	Tom			
Day 9				
Day 10				

Ask children the question, "On which day did your plant grow the most?"

How big was your plant on day ten? Put in order according to size; then place on graph

How many of your beans did not sprout?

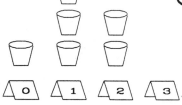

Less than your estimate The same More than your estimate

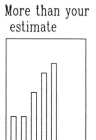

How much did your plant grow since yesterday?

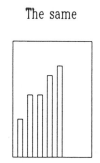

Is your plant still growing?

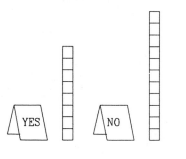

70

BEANS   NAME

① ② ③ ④ ⑤ ⑥ ⑦ ⑧ ⑨ ⑩

Have students create an individual plant graph.

BEANS   NAME

⊗ ② ③ ④ ⑤ ⑥ ⑦ ⑧ ⑨ ⑩

On day 1, ask "Did your plant grow today?"
If there is no growth, students cross out the day.

When growth is visible, have children take a construction paper strip and cut it to match the height of their plants.

BEANS   NAME

⊗⊗⊗⊗ ⑤ ⑥ ⑦ ⑧ ⑨ ⑩

- At the end of the tenth day do the following:

On What Day Did You First See Your Lima Bean Plant?

				Ernie						
				Tim	Rich	Beth				
			Ned	Paul	Alan	Rich			Fern	
			Jim	Sara	Beryl	Noah	Liz		Ben	
Day 1	Day 2	Day 3	Day 4	Day 5	Day 6	Day 7	Day 8	Day 9	Day 10

What is Your Favorite Summer Activity?

June
Grade Level: Primary

TASK ANALYSIS: 6 – Makes symbolic graphs comparing groups
7 – Draws conclusions and makes predictions from data
8 – Colors squares of a grid to correspond to given numbers
9 – Interprets bar graphs as to which shows more and fewer

MATERIALS: Butcher paper graph with heading, names of children on paper strips, xerox one inch grid paper (one per child), crayons.

ORGANIZATION: Whole class activity
Kindergarten/primary: 20 to 30 minutes

PROCEDURE: – Brainstorm for ideas of summer activities.

Comparing Interpreting Recording Graphing	Camping	Swimming	Hiking	Reading

– Ask the children "What is your favorite summer activity?"
– Have children put their names on paper strips and glue under appropriate heading.
– Read and interpret.
– Have children transfer data to grid with crayons.
– Read and interpret.

What Will You Do This Summer?

June
Grade Level: Primary

TASK ANALYSIS: 3 – Makes picture graphs comparing two groups
5 – Records and summarizes data for an event

MATERIALS: Car and house pictures for heading on butcher paper graph, xerox of cars and houses (one per child), glue stick, scissors.

ORGANIZATION: Whole class activity
Kindergarten/primary: 20 to 30 minutes

PROCEDURE: – Make butcher paper graph.

Comparing
Interpreting
Recording
Graphing

What Will You Do This Summer?

– Ask the question "What will you do this summer?"
– Have children cut out appropriate picture and glue under proper heading.
– Read and interpret.

What is Your Favorite Flavor of Ice Cream?

June
Grade Level: Primary

TASK ANALYSIS: 4 – Makes picture graphs comparing three or more groups
5 – Records and summarizes data for an event

MATERIALS: Butcher paper graph, xerox of scoops of ice cream (one per child), glue stick, crayons, scissors

ORGANIZATION: Whole class activity
Kindergarten/primary: 20 to 30 minutes

PROCEDURE: – Make butcher paper graph with three cones.

My Favorite Flavor of Ice Cream

vanilla   chocolate   strawberry

Comparing
Interpreting
Recording
Graphing

– Discuss favorite flavors of ice cream.
– Have children color scoop of ice cream the same color as their favorite flavor.
– Cut out scoop of ice cream.
  Glue to appropriate cone.
– Read and interpret

EXTENSIONS: Make a bar graph
Make ice cream: A good recipe...

1 qt. half and half
4 cups whipping cream
3/4 cup sugar
1 Tbsp. vanilla
1/8 tsp. salt
Mix in ice cream freezer
Eat and enjoy!

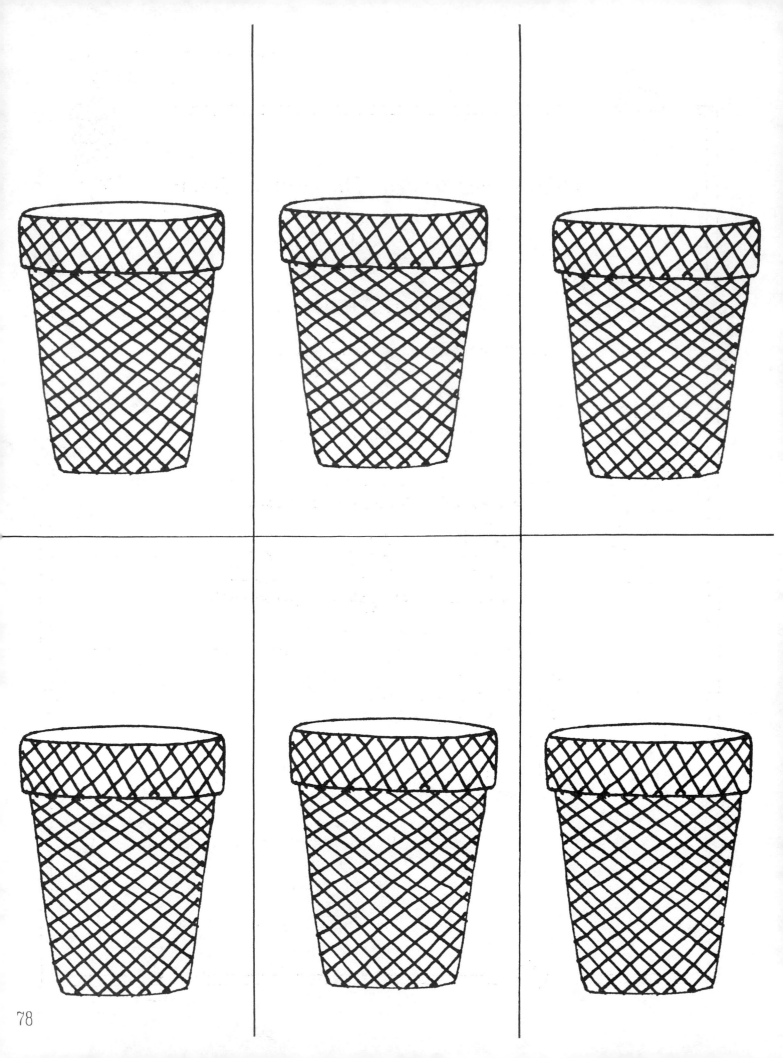

| 11 | Completes and/or constructs bar graphs given data |

A Weighty Problem

Grade Level: Middle

MATERIALS: A bathroom scale, graph paper, rulers, and markers

ORGANIZATION: A whole class activity

PROCEDURE: Have ten students volunteer to be weighed and record their weights on the chalkboard or overhead.

Since some students may be sensitive about their weights it is better to ask for volunteers than to weigh everyone in the class.

Each student should then arrange the weights from least to greatest and create a bar graph representing this information. Discuss the benefits of representing material on a graph rather than on a table. Outline the merits of each method.

This lesson can be extended to include discussion of mean (average), range, median, and mode.

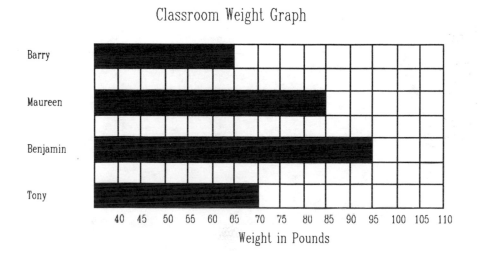

Classroom Weight Graph

Hands On, Inc
2121 Rebild Drive
Solvang, CA 93463

11	Completes and/or constructs bar graphs given data

All the News That's Fit to Graph

Grade Level: Middle

MATERIALS: Newspapers or recordings of newscasts from the radio, graph paper

ORGANIZATION: Groups of two, three, or four students

PROCEDURE: There are various types of news stories. These may include: world news, national news, local news, news of personalities, or human interest stories. This activity asks students to read the front section of a newspaper or listen to a five to ten minute newscast and then generate a graph which depicts they type of news stories reported.

Divide the class into groups and give each group a newspaper (or recording). Give them time to categorize the information into story types. They should include every story in their research.

This information should then be written as a bar graph to share with the class as a whole. Students should include a title as well as the source of the news.

An extension of this lesson would be to have students analyze the number of different types of stories.

Hands On, Inc
2121 Rebild Drive
Solvang, CA 93463

11	Completes and/or constructs bar graphs given data

Clothes Count

Grade Level: Middle

MATERIALS: Pencils, tally sheets (see diagram), and graph paper

ORGANIZATION: Best done in teams of two students.

PROCEDURE: Ask students to predict the most popular clothing color at the school for this particular day. Record student's responses and discuss possible ways of gathering this information. Allow students to gather information during recess.

Divide the class into teams of two students and assign each team a different clothing item to tally:

Team A – skirts
Team B – shirts
Team C – tennis shoes

You may also add items including dresses, sweaters, shorts, shoes, socks, etc. (some duplication may be necessary).

Clothing Color Tally Sheet

Clothing Color	
Red	
Blue	
Tan	
Brown	
Green	

During recess, ask each team to tally their assigned clothing item for twenty students selected at random.

When students return to the room, have each team create a bar graph which depicts their information. Post all completed bar graphs on the board and have students compile all information to decide the most popular color.

Hands On, Inc
2121 Rebild Drive
Solvang, CA 93463

11	Completes and/or constructs bar graphs given data

How Do You Spend Your Time?

Grade Level: Middle

MATERIALS: Graph paper, pencils, and rulers

ORGANIZATION: To be done individually or in teams of two students

PROCEDURE: Ask students to evaluate an average week of their time and estimate the amount of time spent in various activities. Activities might include sleeping, time at school, doing homework, watching T.V., playing, or doing chores.

You may wish to have students evaluate a twenty-four hour period or a week-long time frame.

Once students have gathered their information they should round off the time spent at each activity to half-hour segments and create a bar graph depicting the data.

You may wish to extend this lesson to include averaging the various time segments for the whole class and then creating an "all class" bar graph with the collated data.

Hours Spent at Various Activities

 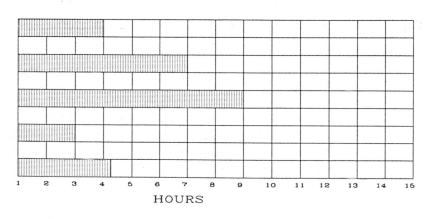

Hands On, Inc
2121 Rebild Drive
Solvang, CA 93463

12	Reads and interprets vertical and horizontal bar graphs

Bar Graph Hunt

Grade Level: Middle

MATERIALS: Newspapers and magazines, especially USA Today, Time, Newsweek, or U.S. News and World Report

ORGANIZATION: Individually or in teams of two

PROCEDURE: Hand out magazines and have students cut out various types of bar graphs. Glue these cutouts on separate sheets of paper and have students create questions about their graphs.

These student generated graph pages can be circulated throughout the room for response and practice by each individual. You may wish to laminate these papers for future use.

Sample Questions:

1. How can you tell which item in the graph is the most popular (most used, most important)?
2. How can you tell which item on the graph is the least popular?
3. How much longer is the longest bar than the shortest? How can you figure this out?
4. What is the value of the most popular (most used, most important) item on the graph?
5. Which item on the graph is used an "average" amount of time? How do you know?
6. What are the benefits of putting this information in graph form?
7. Do you find graphs to be a helpful way of presenting information? Why?

There are several task analysis items involving "reading and interpreting" various types of graphs. The intent of this book is to provide activity based math lessons and in general it is difficult to do reading and interpreting in an activity approach; therefore, we have provided universal lessons for these T.A. items and suggest you use these lessons in conjunction with your basic math series or with duplicated material.

Weather Graphs

Grade Level: Middle

MATERIALS: Copies of five days of weather reports from a newspaper, graph paper

ORGANIZATION: Pairs of students

PROCEDURE: Hand out the information on weather for five days. Ask students to analyze the information and as a team try to decide on information which could be presented as a bar graph.

As a class, discuss each team's selection as to why it will or will not work as a graph. Encourage students to try different approaches with different types of data. There are a variety of acceptable approaches.

Ask each team to plan its graph in terms of the value of the horizontal or vertical axis. Have each team draw its plan on the board. Discuss these plans with the entire class and then complete two or three of the graphs on the chalkboard or overhead.

From Table to Graph

Grade Level: Middle/Upper

MATERIALS: Table presented in appendix C, graph paper

ORGANIZATION: Can be done individually or in groups of two, three, or four

PROCEDURE: Hand out copies of appendix C to each student or group. Tell them that they will be constructing a bar graph which depicts this information.

Essential elements to explain include: a title for the graph, labeling of the vertical and horizontal axis, neatness, and accurate transfer of data.

You may wish to work with the students in a step by step procedure or simply present the concept and have each group work together to arrive upon a finished product.

Hands On, Inc
2121 Rebild Drive
Solvang, CA 93463

13	Generates original data and constructs a bar graph

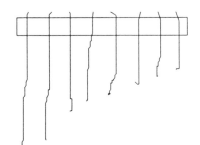

Soleful String

Grade Level: Middle

MATERIALS: Scissors, rulers, string, masking tape, and a chalkboard or display board

ORGANIZATION: Students work in pairs or in teams of three

PROCEDURE: Ask each student to cut a piece of string the estimated length of his shoe. The teacher may wish to demonstrate to help standardize the procedure. Have students check their guesses.

Next, have students cut another piece of string to the actual length of their shoes (heel to toe), measure the string (in inches or centimeters), and place their lengths of string on the chalkboard with tape as shown. You might also discuss the difference between their guesses and the actual size.

When all students have done this, ask students to come to the board and rearrange the strings from shortest to longest. Have other students group the strings into 5 inch strings, 6 inch strings, etc.

Students should then graph this information on 1″ graph paper.

Extensions may include discussion of mode, median, range, and mean (average). You may also wish to extend this to a double bar chart by measuring shoes from side to side or by comparing younger vs. older children.

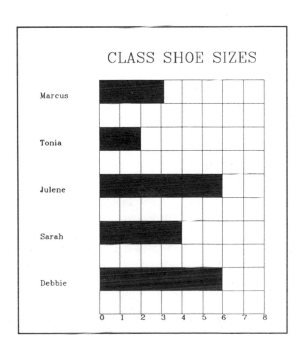

Hands On, Inc
2121 Rebild Drive
Solvang, CA 93463

13	Generates original data and constructs a bar graph

Home Hunting

Grade Level: Middle/Upper

MATERIALS: Graph paper, rulers

ORGANIZATION: Can be done in teams but is best done individually

PROCEDURE: As a homework assignment, have each student complete the following questionnaire:

How many chairs are in each room of your house?
    Kitchen:
    Dining Room:
    Living Room:
    Bedroom 1:
    Bedroom 2:
    Bedroom 3:
    All Others:
How many windows are in each room of your house?
    Kitchen:
    Living Room:
    Dining Room:
    Bedroom 1:
    Bedroom 2:
    Bedroom 3:
    All Others:

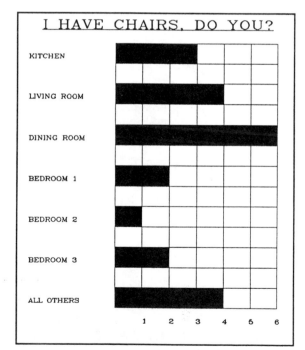

Have students return to class the next day and make two bar graphs which show this information. Bar graphs should have titles and labels. You may wish to have students do one graph vertically and the second horizontally.

As an extension you may ask students to do a composite graph of this information for the entire class.

Hands On, Inc
2121 Rebild Drive
Solvang, CA 93463

| 13 | Generates original data and constructs a bar graph |

Food, Dude!

Grade Level: Middle/Upper

MATERIALS: Tally sheets, graph paper, markers, and rulers

ORGANIZATION: Teams of three or four students

PROCEDURE: This is a two day project with set up and information gathering on the first day and organization and graphing on the second.

Have students circulate during the lunch hour collecting data about the types of lunches children are eating. Each team can take a different aspect of lunch fare with items such as types of sandwiches, number of sandwiches, type of beverage, type of dessert, type of fruit, and what students eat and don't eat.

After this information is gathered, students will need to collate information from each team member and figure out percentages in order to complete a bar graph depicting the data.

You may request that students prepare an oral presentation or written report explaining the experiment and the results. You might also have students prepare a newsletter to parents explaining the results. There is also the obvious tie to a unit on food groups and nutrition.

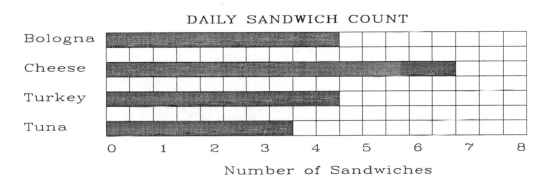

DAILY SANDWICH COUNT

Hands On, Inc
2121 Rebild Drive
Solvang, CA 93463

13	Generates original data and constructs a bar graph

On What Day Were You Born?

Grade Level: Middle/Upper

MATERIALS: Graph paper, rulers, markers (optional), use of a perpetual calendar (in an almanac, data watches, and some computer programs)

ORGANIZATION: Teams of two, three, or four students

PROCEDURE: Have each team collect data from all other students regarding day of the week, date, month, and year of birth -- each team gathering the same set of data.

Much time will be spent in demonstrating to students the methods for using per-petual calendars since this concept will probably be new to them.

Each team should decide upon the best method of converting this data.

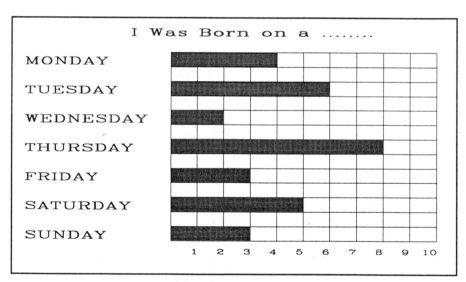

Taking the raw data, each team should sort, organize, and prepare a bar graph depicting the actual day (of the week) of birth -- each group's information should match.

Hands On, Inc
2121 Rebild Drive
Solvang, CA 93463

14	Reads and interprets pictographs

Pictograph Hunt

Grade Level: Middle

MATERIALS: Newspapers and magazines, especially USA Today, Time, Newsweek, or U.S. News and World Report

ORGANIZATION: Individually or in teams of two

PROCEDURE: Hand out magazines and have students cut out various types of pictographs. Glue these cutouts on separate sheets of paper and have students create questions about their graphs.

These student generated graph pages can be circulated throughout the room for response and practice by each individual. You may wish to laminate these papers for future use.

Sample Questions:

1. How can you tell which item in the graph is the most popular (most used, most important)?
2. How can you tell which item on the graph is the least popular?
3. How much longer is the longest bar than the shortest? How can you figure this out?
4. What is the value of the most popular (most used, most important) item on the graph?
5. Which item on the graph is used an "average" amount of time? How do you know?
6. What are the benefits of putting this information in graph form?
7. Do you find graphs to be a helpful way of presenting information? Why?

There are several task analysis items involving "reading and interpreting" various types of graphs. The intent of this book is to provide activity based math lessons and in general it is difficult to do reading and interpreting in an activity approach; therefore, we have provided universal lessons for these T.A. items and suggest you use these lessons in conjunction with your basic math series or with duplicated material.

Weather Graphs

Grade Level: Middle

MATERIALS: Copies of five days of weather reports from a newspaper, graph paper

ORGANIZATION: Pairs of students

PROCEDURE: Hand out the information on weather for five days. Ask students to analyze the information and as a team try to decide on information which could be presented as a pictograph.

As a class, discuss each team's selection as to why it will or will not work as a graph. Encourage students to try different approaches with different types of data. There are a variety of acceptable approaches.

Ask each team to plan its graph in terms of the value of the horizontal or vertical axis. Have each team draw its plan on the board. Discuss these plans with the entire class and then complete two or three of the graphs on the chalkboard or overhead.

From Table to Graph

Grade Level: Middle/Upper

MATERIALS: Table presented in appendix C, graph paper

ORGANIZATION: Can be done individually or in groups of two, three, or four

PROCEDURE: Hand out copies of appendix C to each student or group. Tell them that they will be constructing a pictograph which depicts this information.

Essential elements to explain include: a title for the graph, labeling of the vertical and horizontal axis, a legend telling the value of each picture, neatness, and accurate transfer of data.

You may wish to work with the students in a step by step procedure or simply present the concept and have each group work together to arrive upon a finished product.

 Hands On, Inc
2121 Rebild Drive
Solvang, CA 93463

15	Constructs a pictograph given data

Counting the Class

Grade Level: Middle

MATERIALS: Construction paper, glue, scissors

ORGANIZATION: Groups of two, three, or four students

PROCEDURE: This is a very basic lesson in constructing pictographs. Students will be using a picture to represent 2 or 3 students. This representation is a very difficult concept for young children to understand.

Begin the lesson by explaining the purpose of a pictograph. Tell students that they will be creating a pictograph showing the number of boys and girls in the classroom.

For their first experience, you will need to guide them step by step: counting students, discussing the possibility of cutting out a symbol for each student or letting one symbol count for 2 or 3 students, explaining that a fractional part of a picture might need to be used to represent an uneven number of students, and finally the set-up of the graph itself.

Once students have completed this graph, you may wish to divide them into groups to complete a pictograph showing the types of shoes worn by classmates, colors of clothing, or perhaps color and types of buttons. You may also choose to increase the sample by having students go to other classes to gather information on the number of boys and girls or shoe styles.

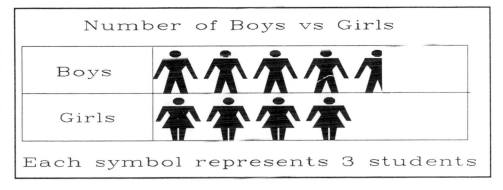

15	Constructs a pictograph given data

Pages Count!

Grade Level: Middle +

MATERIALS: Each student will need to have access to five different books at his desk.

ORGANIZATION: To be done individually

PROCEDURE: Tell students that they are going to make a pictograph representing the number of pages in five of their books. Have them get five books from their desks or from some source and go through and record the number of pages in each book.

Given this information, have students decide upon a representative symbol and the number of pages each symbol will represent in order to complete a pictograph. Have students draw the symbols or cut them out of construction paper for completion of the graph.

You may wish to extend this lesson to include a discussion of range, median, mean (average) and mode using the data which students have gathered.

BOOK TITLE	PAGES PER BOOK
Science in Action	● ● ● ●
Latin America Today	● ● ●
Happy Trails	● ● ●
Math Around Us	● ●
Your Language	● ● ● ● ●

● = 50 pages

 Hands On, Inc
2121 Rebild Drive
Solvang, CA 93463

15	Constructs a pictograph given data

"Ad" This to Your Magazine
Grade Level: Middle

MATERIALS: Magazines (one for each student), glue, construction paper, and scissors

ORGANIZATION: Individually or in teams of two.

PROCEDURE: This lesson combines graphing skills with logic. Students will be categorizing magazine advertising into main topics.

Let each child select a magazine to analyze. It is best if magazines are no more than 50 pages. If larger magazines are used, you may wish to have students use the front portion only.

As they look through the pages, they should tally the number of advertisements for various types of products. This will involve the logic of classifying similar products into a main heading (i.e. health care products might include toothpaste, deodorant, shampoo, etc.).

Once they have tallied the various products, they should cut out a symbolic picture (a toothpaste tube, for example) to represent a given number of health care products.

The graph shown depicts an advertising breakdown of 11 health care products, 5 cosmetic products, 13 apparel advertisements, and 2 advertisement for foreign travel.

MAGAZINE ADVERTISING SUBJECTS

Health Care	🪥 🪥 🪥 🪥 🪥 🪥
Cosmetics	🧴 🧴 🧴
Travel	✈️
Apparel	👚 👚 👚 👚 👚 👚 👚

Each picture represents two advertisements

Hands On, Inc
2121 Rebild Drive
Solvang, CA 93463

15	Constructs a pictograph given data

Gaining Favor

Grade Level: Upper

MATERIALS: Graph paper, tally sheets, rulers, and markers.

ORGANIZATION: To be done as a whole class activity or in teams of two students

PROCEDURE: Students take turns interviewing their partners on topics such as their five favorite TV shows, the seven most popular bands, amount of time per day spent reading, doing homework, watching TV, at hobbies, etc. Collection of data might be teacher directed or can be done in teams if the class is able to handle the activity.

Once data collection is complete, work with students to construct a series of pictographs which depicts the information. You may also wish to model completion of one or two graphs and then divide students in groups of four to complete graphs on their own.

FAVORITE TV SHOWS

Head of the Class	▭ ▭ ▭ ▭
Who's the Boss	▭ ▭ ▭
Growing Pains	▭ ▭ ▭
Family Ties	▭ ▭ ▭ ▭ ▭
Newhart	▭ ▭

 Hands On, Inc
2121 Rebild Drive
Solvang, CA 93463

16	Generates data and constructs original pictographs

Either/Or Nothing More

Grade Level: Middle

MATERIALS: Pencil and paper (construction paper, scissors, and glue if you wish students to make a more colorful display)

ORGANIZATION: To be done individually

PROCEDURE: Have each student think of a question which is a choice between two popular selections such as:

Which do you prefer? Pepsi or Coke; Cats or Dogs, Pizza or Hamburgers; etc.

Once each student has created a question, have him post the choice on a piece of construction paper and allow students to circulate and write their initials under their choices. This will take some time as they will be voting for an item for each student.

WHICH DO YOU PREFER?

PEPSI	* * * * * * * *
COKE	* * * *
CATS	* * * * * * * * * *
DOGS	* * * * *
PIZZA	* * * * * * * * * * * * * * *
HAMBURGERS	* * * * * * * * *

* Represents 2 students

Once this raw data has been collected, each result is represented by a pictograph using an appropriate scale (such as one picture for five votes). Each student makes his own pictograph with pencil and paper or by cutting out appropriate symbols and gluing them on paper.

Students should be reminded to include a title and key for the graph.

Extensions include surveying parents on the same questions or interviewing a larger sample such as other math classes. You may have students present their information orally to the class. This information might also be presented as a double bar graph.

Hands On, Inc
2121 Rebild Drive
Solvang, CA 93463

16	Generates data and constructs original pictographs

Shock Value

Grade Level: Middle

MATERIALS: A tally sheet created by students in class or by teacher which lists the various types of electronic appliances which students may have at home.

ORGANIZATION: To be done individually

PROCEDURE: Students will be collecting information on the number of appliances at home and then collating this data in a classwide pictograph.

As a homework assignment, have each student scour the house writing a list of each appliance which plugs in. They might include hair dryers, TV's, washing machines, computers, food processors, etc.

Have students decide as a class the best method of collating all of this information. The cooperative learning aspect of this assignment is an added benefit.

Once students have gathered all of the information, have the class create a pictograph which depicts all of the appliances owned.

As an extension you might want to research cost of all of these items to establish a means of extending this lesson to other math strands or other curricular areas..

ITEM	NUMBER FOUND
Hairdryer	⚠ ⚠ ⚠
Television	⚠ ⚠ ⚠ ⚠
Washing Machine	⚠ ⚠ ⚠
Computer	⚠
Food Processor	⚠
Clock Radio	⚠ ⚠
Stereo	⚠ ⚠ ⚠
Walkman	⚠ ⚠ ⚠ ⚠
Lamps	⚠ ⚠ ⚠ ⚠ ⚠

⚠ = 4 APPLIANCES

16	Generates data and constructs original pictographs

Tally Ho!

Grade Level: Middle

MATERIALS: Pencils, tally paper, scissors, construction paper, glue

ORGANIZATION: To be done individually or in teams of two, three, or four.

PROCEDURE: Each team selects a class or school wide object to count and categorize. This might include the number and variety of windows in the school, the number and types of desks, the variety of types and colors of books, pencils or pens in the room, or the number and types of trees surrounding the school.

Tell students that they are going to gather this information and display it in a pictograph.

When students have finished collecting data, they should decide upon a suitable symbol for their graph and the number of items each symbol will represent. Students should be given time to complete their pictographs.

Using cutouts and gluing them onto construction paper is a more satisfactory method than having students simply draw the graph.

SCHOOL TALLY PICTOGRAPH

WINDOWS	# # # # # #
DESKS	# # #
CHAIRS	# # # # # # #
PENCILS	# # # # # # # # # # # # #
PENS	# # # # # # #

= 10 ITEMS

 Hands On, Inc
2121 Rebild Drive
Solvang, CA 93463

| 16 | Generates data and constructs original pictographs |

Sing! Sing! Sing!

Grade Level: Middle

MATERIALS: Two or three tape recorded songs on cassettes (the teacher should be at least vaguely familiar with the selections), tally sheets

ORGANIZATION: Teams of two students

PROCEDURE: Tell students that they are going to analyze the repetition of words in popular music. Ask these questions and record student estimates on the chalkboard:

How many words do you think there are in an average song?
How many times does the singer repeat certain words and phrases (on the average)?
In a given song, how many times would the words "the" or "a/an" appear?

After recording these responses, assign each team a certain word, phrase, count of words, or melody or rhythm to listen for as you play a tape. For example, TEAM 1 might count the use of "the" in the tape. TEAM 2 might count the use of the word "love." TEAM 3 might count the total number of words used in the song, etc. You will undoubtedly have to play the song two or three times to gather the data.

Given this information, have students compile information and create a class graph which depicts the frequency of word usage.

As a homework assignment, have each student select a song and conduct the same experiment. The results will amaze the students and the teacher as well!

The	● ● ● ●
A/An	● ● ● ● ●
Love	● ●
Uh uh	● ●
You	● ● ●
The Title	● ● ● ●
I	● ●
Total Words	● ● ● ● ● ● ● ● ● ● ●

● = Two uses of the word

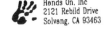 Hands On, Inc
2121 Rebild Drive
Solvang, CA 93463

| 17 | Reads, interprets, and constructs tables for organizing data and solving simple problems |

Two For a Nickel

Grade Level: Middle

MATERIALS: A large supply of new, unsharpened pencils; play money nickels, bottle caps, or something to represent nickels.

ORGANIZATION: Groups of four

PROCEDURE: Establish groups each having a table or flat working area.

Pass out pencils and "nickels" and tell students that they should imagine themselves to be shopkeepers and that two pencils cost one nickel. Have each student create this physical arrangement. Extend this to have students arrange four pencils and two nickels, then six pencils and three nickels.

Pencils	Nickels
2	1
4	2
6	3
8	4

Ask if students can think of any way to show this "pattern" on paper. They may draw pictures to begin with, but eventually,they should arrive upon a table to solve the problem.

Have students complete the table through 12 pencils and ask if they see a pattern to help them solve a problem of the cost of 20, 30, or 40 pencils.

Extend the lesson by changing the cost for pencils -- one pencil for two nickels, three pencils for a nickel, and other combinations. Students should eventually verbalize that a table is a structured and simple means of representing patterns and proportion.

Hands On, Inc
2121 Rebild Drive
Solvang, CA 93463

17	Reads, interprets, and constructs tables for organizing data and solving simple problems

Take Your Pick

Grade Level: Middle

MATERIALS: Boxes of toothpicks

ORGANIZATION: Students may work individually or in pairs

PROCEDURE: Give each team a stack of 30 to 40 toothpicks and tell them that they are going to create a table which demonstrates the number of toothpicks needed to make various numbers of squares.

Begin by asking students to make one square and ask how many toothpicks were used (4).

To make a second square adjoining the first, students will need three more toothpicks and so on. Have students make a table (a sample is given) and complete it up through ten boxes.

Discuss with students the advantage of organizing information in tabular form and have them predict the number of toothpicks needed to form 100 boxes (and so on).

Have students choose a different shape and complete a second table. The shape may be a triangle, rectangle, non-connected shapes. The purpose of this lesson is to have students identify tables as a form of organizing, predicting and problem solving.

# of ☐	Toothpicks
1	4
2	7
3	10
4	13
5	16

 Hands On, Inc
2121 Rebild Drive
Solvang, CA 93463

17	Reads, interprets, and constructs tables for organizing data and solving simple problems

A Hit 'n' Miss Approach

Grade Level: Upper

MATERIALS: Hammer, nails of various sizes (6d,8d,10d,etc), blocks of scrap wood, paper, and pencil -- A stump of some description is an excellent (and quiet) nail receiver.

ORGANIZATION: Teams of two

PROCEDURE: Give materials to each team. One team member will hammer while the other counts the number of strokes it takes to hammer the nail down completely.

The counting team member records the number of hits and misses. Students take turns hammering and recording the results of different nails and blocks of wood.

Using the information gathered on tally sheets, students construct a table of their own data. The table may be set up in a variety of forms (note sample).

As an extension you may ask students to create proba-

Nail Size	# of Hits	# of Misses
6d	9	2
8d	14	5
10d	19	9
12d	25	15
14d	33	21

bilities as to the number of misses as related to the sizes of the nails; or to find ratios between the nail size and the number of hits taken to drive in the nail.

Hands On, Inc
2121 Rebild Drive
Solvang, CA 93463

17	Reads, interprets, and constructs tables for organizing data and solving simple problems

A Yen for Dollars

Grade Level: Upper

MATERIALS: Almanacs, encyclopedias, financial pages from various newspapers, and other sources of information about currency from foreign countries.

ORGANIZATION: Can be done individually or in groups of two, three, or four students

PROCEDURE: Tell students that they are going to take an imaginary trip to five countries of their choice. .Give them $200.00 to spend on various items and have them write a list of their purchases and the cost of each item. Guide students to buy anywhere from five to ten different souvenirs.

Once students have done this explain that in foreign countries, U.S. currency is not accepted -- it must be converted to the currency of that country. Have students create a table similar to the one pictured and have them complete the currency conversion from U.S. to foreign dollars.

Students should then rewrite their lists of purchases figuring the amount of "foreign" dollars each item would cost.

COUNTRY	AMOUNT EQUAL TO $1.00 U.S.	ITEM PURCHASED	TOTAL FOREIGN COST	AMERICAN COST
Italy				
Germany				
Spain				
England				
Portugal				

Hands On, Inc
2121 Rebild Drive
Solvang, CA 93463

18	Manipulates real objects, bar graphs, and pictographs to find average

What's in a Name?

Grade Level: Middle

MATERIALS: Graph paper, scissors, glue

ORGANIZATION: Most effectively done individually but can be done in groups of two, three, or four students.

PROCEDURE: Have students make a list of the names of relatives or classmates. On the graph paper write the names -- one letter per square. Cut out each name (in strips) and assemble them from shortest to longest.

To find the MEAN, have students cut one letter at a time from the end of the longer names and add them to the ends of the shorter names (as shown in the diagram) until all names are the same length (approximately). Glue the "averaged" names on a sheet of paper.

Discuss the meaning of the word MEAN (average) with students. Have students verbalize the process of obtaining the average. What did they do to reach this solution? How did they know when to stop cutting letters?

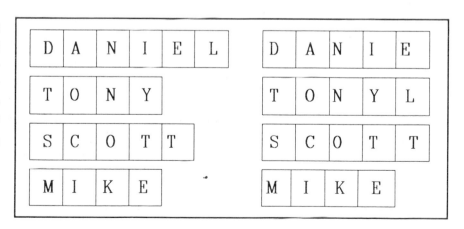

18	Manipulates real objects, bar graphs, or pictographs to find average

Just an Average Shoe

Grade Level: Middle

MATERIALS: Paper, pencil, graph paper, colored pens, and a class list

ORGANIZATION: Groups of four

PROCEDURE: Gather data of shoe sizes, one for boys and one for girls. Construct two bar graphs, one for the boys' shoe sizes and one for girls'. Using this graphed information, have students find the mean (average) shoe size for the boys and and the girls.

Discuss the methods used by each group to decide upon the mean. This may lead to a discovery or affirmation of the standard method of adding all sizes and dividing by the amount of number added.

Transfer this information to the board or an overhead projector to discuss median, range, and mode of the shoe size data.

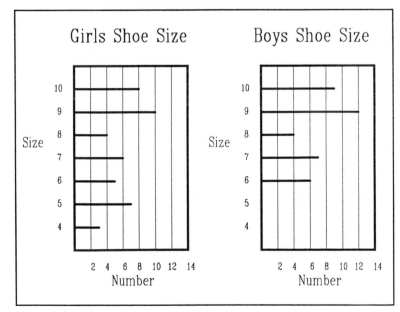

Hands On, Inc
2121 Rebild Drive
Solvang, CA 93463

18	Manipulates real objects, bar graphs, and pictographs to find average

Let it Roll

Grade Level: Middle

MATERIALS: Dice for each group and graph paper

ORGANIZATION: Divide students into teams of two

PROCEDURE: Tell students that they are to try to predict the mean (average) number of twenty-rolls of the dice. Most students will surmise that the mean should be 6 or 7, but they should try to come up with a rationale as to why this is true.

One team member rolls the dice while the other person records the results on a tally sheet. At least twenty rolls should be included in the sample. Team members can trade turns so that each gets an opportunity to participate.

Each team should then convert this tally sheet into a bar graph, and from the graph students should try to prove that their prediction was true. Allow students to discover ways of manipulating the graph to create a mean or average.

You may wish to have students prepare a paragraph explaining their methods and why the result should be 6 or 7. They should also explain why 6 or 7 may not be the proven answer (due to a small sample size).

TALLY SHEET	
1	|
2	||
3	||
4	|||
5	||||
6	|||||
7	|||||
8	|||
9	||||
10	|||
11	||
12	|

Hands On, Inc
2121 Rebild Drive
Solvang, CA 93463

18	Manipulates real objects, bar graphs, or pictographs to find average

Cut Ups

Grade Level: Middle

MATERIALS: Scissors, string, and masking tape

ORGANIZATION: Groups of five, six, or seven students

PROCEDURE: Students will attempt to find their average height by measuring one another with string, then cutting and taping pieces of string together to create equal length pieces.

Give each group a long length of string. Working in pairs, measure one another by cutting a string length to each student's height. Once each person has a string, lay them out on the floor and begin cutting lengths from the longest string and taping them to the shortest.

Repeat this process until all string lengths are approximately equal.

Discuss with students other methods they might use to determine the average or mean. There are a variety of methods. Allow students the opportunity to be creative.

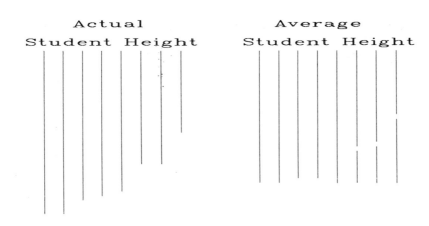

Actual
Student Height

Average
Student Height

Hands On, Inc
2121 Rebild Drive
Solvang, CA 93463

19	Computes average (mean) from given data

Using Your Bean

Grade Level: Middle

MATERIALS: A large bag of beans

ORGANIZATION: Individuals or groups of any size

PROCEDURE: Each student takes a handful of beans and counts the total (groups record the amount for each individual). Each individual's amount is then written on the board so the entire class can see the totals.

Students are then asked to determine the type of information they can gather from the various numbers. They should be directed to notice: the least amount of beans in a handful (low end of the range), the largest amount of beans in a handful (high end), and the most common number of beans. Teaching range and mode is not the intent of the lesson, but it is good to point out this variation for future reference.

Ask each group to develop a method by which every student in the class will have the same number of beans. Each group should explain its method.

One of the most obvious responses will be to have students exchange beans with all members of the class until each member has the same amount (plus or minus one). You may choose to use other methods as well to encourage the students' independent thinking.

When students have completed the task of evenly distributing beans, write the number of beans each student has on the board. Demonstrate that this number can also be reached by adding the various handful amounts and then dividing by the total number of students.

Tell students that this number is the mean or average.

Encourage students to redistribute beans and see if they can solve the problem both physically (by distribution) and by mathematical operation.

19	Computes average (mean) from given data

Numbers Can Be Mean

Grade Level: Upper

MATERIALS: A phone book, atlas, or other publications which contain series of numbers

ORGANIZATION: Teams of two, three, or four students.

PROCEDURE: Give each group a source for finding a list of numbers. These can be phone numbers, populations, addresses, etc. The numbers should be random.

242−4396
242−6039
242−1846
242−0019
242−5992
242−0192
242−0873
242−7641
242−0858

Ask students to select a set of twenty-five numbers and write down the digits which appear in the "one's place." Have each group estimate the average of these numbers.

By using the standard approach of adding the numbers and dividing by twenty-five, see how close their estimations were.

Generally speaking, the average of any twenty-five random numbers between 0 and 9 should be 4.5. On the chalkboard, total all group numbers and see if this occurs.

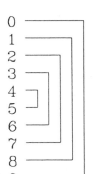

Have students complete the same procedure for the numbers in the tens place (average should be 50) and the hundreds place (average should be 500).

Discuss the reasons for the average continually being near this midpoint (see diagram).

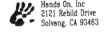
Hands On, Inc
2121 Rebild Drive
Solvang, CA 93463

19	Computes average (mean) from given data

Flight Simulator

Grade Level: Upper

MATERIALS: Graph paper, rulers, paper, string

ORGANIZATION: Students are grouped in pairs

PROCEDURE: In pairs, students are given a series of experiments to perform and record results:

1. Tossing wads of paper to each other (or an agreed upon target) and tallying catches vs. total number of attempts.
2. Using paper airplanes, students see if they can land within three feet of their partner.
3. Using the airplanes, string, and rulers, students measure the distance of their flights.

Results of the various activities are then tabulated and averages calculated.

An extra step might be added to convert this information into bar graph form; however, this approach is used in several of the preceding lessons.

As an extension you might introduce the concept of representing probability as a fraction.

Paper Toss

Attempts	Catches
10	6
10	8
10	10
10	9

Airplanes for Accuracy

Attempts	w/i 3 ft.
10	2
10	4
10	5
10	4

Airplanes for Distance

Flight	Distance
1	10'
2	14'
3	13'
4	19'

Hands On, Inc
2121 Rebild Drive
Solvang, CA 93463

19	Computes average (mean) from given data

They're Playing Our Commercial!

Grade Level: Upper

MATERIALS: Tally sheets broken into minutes over a fifteen to thirty minute period, recorded radio shows, tape recorders or transistor radios.

ORGANIZATION: Teams of two, three, or four students

PROCEDURE: Ask students to make a guess as to the amount of time a radio station spends doing commercials versus playing music. As groups, they will be analyzing some of this data and then compiling an average ratio.

Hand out tapes to each group and allow fifteen to thirty minutes of listening time. Have students tally the music time and commercial time (count talk time as commercials). The attached chart shows a sample result.

Once students have gathered the information, have each group record result on the chalkboard. Allow each group time to find the average time of music and of commercials.

Students may use a variety of methods to compute this information. Upon completion of the project, give each group time to explain its method of computing the average.

As an extension you might ask children to do the same experiment at home with television programs.

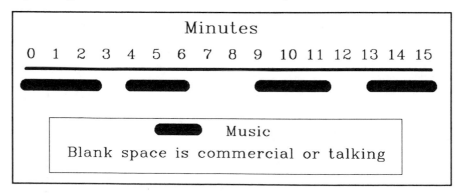

 Hands On, Inc
2121 Rebild Drive
Solvang, CA 93463

20	Identifies range, median and mode given data

What's in a Name?-- Part II

Grade Level: Upper

MATERIALS: Graph paper, scissors, glue

ORGANIZATION: Most effectively done individually, but can be done in groups of two, three, or four students.

PROCEDURE: This activity is an extension of <u>What's in a Name</u> in Task Analysis 20. You may wish to combine the lessons.

Have students make a list of the names of relatives or classmates. On the graph paper write the names – one letter per square. Next to each name write the number of letters in each name in a separate box.

Cut out each name (in strips) and assemble them from shortest to longest.

Have the students find the mean (average) as described in <u>What's in a Name</u> (T.A. 20).

To find the median, have students locate the name in the **middle** of the ordered names (if there is an even amount of names, the two center-most numbers should be added together and divided by 2 to find the median).

Have students discuss how the median and mean differ.

Place all numbers in a graph format. The number which occurs most often is the mode. If there is no number which occurs more than once, there is no mode.

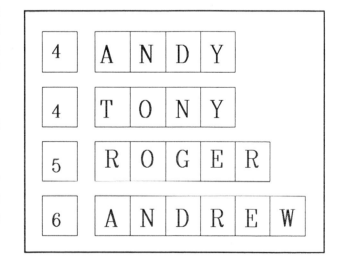

20	Identifies range, median, and mode given data

The Price is Right

Grade Level: Upper

MATERIALS: Various advertisements from newspapers, sale brochures, or announcements; scissors, construction paper, and glue.

ORGANIZATION: Can be done individually or in teams of students.

PROCEDURE: Each student or team should select a product or type of product to research. This may be VCR's, toothpaste, computers, soda types, cereal, etc.

Tell students that they will be doing research on this product and its price in various sources. It would be beneficial to find ten prices for their product.

Once this information is found students should cut out ads and place them in order from lowest price to highest. Given this list, students should identify the RANGE of prices, the MEDIAN and MEAN price, and the MODE if there is one.

The ads can be glued to construction paper and displayed to instigate more discussion. This discussion might include whether the cheapest item is always the best deal or if quality is more important.

RED DOT GOLF BALL PRICES	
Sav On	4.98
Thrifty	5.05
Big Five	4.70
K-Mart	5.25
Oshman's	4.95
Sports World	4.98
Target	5.25
Penney's	5.10
Sears	5.20
Long's	4.98

Hands On, Inc
2121 Rebild Drive
Solvang, CA 93463

20	Identifies range, median and mode given data

Newsworthy Words

Grade Level: Upper

MATERIALS: Newspapers, paper, pencils and rulers measuring to at least 1/8"

ORGANIZATION: To be done individually or in teams

PROCEDURE: Ask students to guess the average length of words in a newspaper. This can be either the actual linear length or the number of letters in each word. Ask them to also predict the range of length (or number of letters) in a given article. You might also predict the median.

Give each individual or group a short excerpt from a newspaper and have students devise a method by which they will find the range, median and mode.

The groups should keep the work to explain both their findings and methods to the class.

Have the class discuss any differences in findings or in methods of collecting data. You may choose to have them collate the information for each group and find a classwide average (mean) from the data they have collected.

A fascinating extension to this activity is to have students experiment with the Fry Readability Scale which takes the number of sentences and number of syllables per 100 word section and establishes a reading ability level.

Candidates Flexing Political Muscle
By Bill Smoot

UPI The Super Tuesday primaries are at hand and it is becoming difficult to tell whether the candidates are running for president of the United States or the toughest guy on the beach.

With issues having failed to capture much attention and time being short, they have turned to their political muscle; each is selling himself as tougher, flintier of character and a greater leader of men than his rivals...

 Hands On, Inc
2121 Rebild Drive
Solvang, CA 93463

20	Identifies range, median, and mode given data

A Square for a Square

Grade Level: Upper

MATERIALS: Strips of graph paper 21 squares long and large enough for a marker of some kind, markers, a die for each child.

ORGANIZATION: Teams of two students

PROCEDURE: Ask children to predict which number would occur most often if they were to roll the die twenty times. Elicit the response that the rolls should spread evenly from one to six. Discuss the fact that the RANGE of rolls would be 1 – 6.

Tell students that they are going to play a dice game called "Squares." The rules for this game are as follows. Each player attempts to move the marker off the end of his side of the graph paper strip. Begin the game by placing the marker in the center square. Player 1 rolls and moves the marker towards him that number of squares. Player 2 then rolls and moves the marker back in his direction (towards the center square or past it). Player 1 rolls and the game continues in this manner. Continue play until each player has had four turns.

Center Start Here

10	9	8	7	6	5	4	3	2	1	0	1	2	3	4	5	6	7	8	9	10

Roll 1 – Student 1 rolls a 3

10	9	8	7	6	5	4	3	2	1	0	1	2	3	4	5	6	7	8	9	10

Roll 1 – Student 2 rolls a 5

10	9	8	7	6	5	4	3	2	1	0	1	2	3	4	5	6	7	8	9	10

The RANGE in the span of squares the marker moved in four turns.
In the two moves shown, the range would be as marked

Occasionally there will be a winner in four rolls, but more typically the marker will be very near the center square. Explain to the children that the center square represents the median, the squares from 1 to 10 on each side of the middle square represent the range of the game.

Ask students if the marker ever landed on the same square (a second time). This would represent the Mode of the game.

In this activity, the concept of range, median and mode are more abstract, but the focus of seeing the center square as the middle of the game will help students to understand the concept of median.

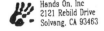

Hands On, Inc
2121 Rebild Drive
Solvang, CA 93463

21	Writes fractions to represent probability

The E's Have It!

Grade Level: Upper

MATERIALS: Paper and pencil, rulers, sources for paragraphs of written material.

ORGANIZATION: Can be done individually or in groups.

PROCEDURE: Have each student write the alphabet down the left hand column of a piece of paper. Have them estimate the number of each letter that will appear in a given paragraph and record their guesses.

Have them select a paragraph of twenty-five to fifty words. Go through and tally the letters. Upon completion they will have a simple pictograph with a one to one correspondence.

A variety of directions can be taken at this point. If students are divided into groups, each group may take a different aspect. Some ideas include:

1. Create a fractional representation of the probability of vowel use vs. consonants.

2. Create a fractional representation of the probability of each letter. For example, if there were 250 letters used and 25 of these were E's, the probability represented as a fraction would be 1/10.

3. Create a fractional representation of capital letters vs. lower case letters.

4. Create a fractional representation letters A-M vs. letters N-Z.

This activity can be extended to making circle graphs of this information.

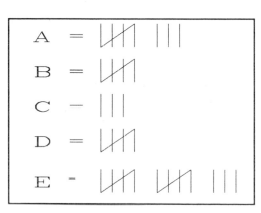

Hands On, Inc
2121 Rebild Drive
Solvang, CA 93463

21	Writes fractions to represent probability

Crash Landing

Grade Level: Upper

MATERIALS: One paper cup of the same size for each group, tally sheets, pencils

ORGANIZATION: Teams of two or three students

PROCEDURE: Ask students to describe the different ways a cup would land if it were dropped (standing up, upside down, or on its side). Ask students to predict how many times each outcome will occur in twenty trials . They should write their predictions.

Use a standardized method for all class members such as standing and dropping the cup on the table or on the floor and tally the way it lands. Using this information, have students predict the outcome if the cup were dropped 40, 60 or 80 times.

Have students convert this information into fractional form (a sample outcome is given).
The probability of the cup landing in standing position is: 1/15
The probablity of the cup landing on its side is: 2/3
The probability of the cup landing upside down is: 1/10

Students will have difficulty making this conversion, and you will need to demonstrate the process before they begin working on this aspect of the lesson.

Once students have created fractional problems they may want to retest for accuracy. Using the sample given above, students might drop the cup fifteen times and see if it lands in a standing position one time; drop the cup three times and see it it lands on its side two times, etc.

Standing up	Upside Down	On Side
|||	||||	⅏ ⅏ ||

Hands On, Inc
2121 Rebild Drive
Solvang, CA 93463

21	Writes fractions to represent probability

A Chip Off the Old Block

Grade Level: Upper

MATERIALS: Bags with a given number of chips of two different colors for each group.

ORGANIZATION: To be done in teams of two, three, or four students.

PROCEDURE: Give each team a bag with FIVE white and FIVE blue chips, but do not allow them to look inside. Tell students that their ultimate goal is to remove chips and return them and then to predict the probability of either color chip appearing.

Students begin by pulling a chip, recording the color, and returning the chip to the bag. The bag is then shaken to randomize the results. Have students do this ten times and then estimate the number of each color in the bag. Continue this process twenty, thirty, or forty times -- enough for students to feel they have an accurate sample.

sample 1		sample 2		sample 3	
White Chips	|||||	White Chips	|||||	White Chips	||||||
Blue Chips	||||||	Blue Chips	||||||	Blue Chips	||||

Based upon this information have students write the probabilities of pulling a white chip and of pulling a blue chip in fractional form.

It is good to start this activity with five chips of each color in each bag. You may wish to prepare a second set of bags with a ratio of 3 to 7. Students will find that if all teams write the results of the sample on the board, the sample becomes larger and more accurate.

Hands On, Inc
2121 Rebild Drive
Solvang, CA 93463

21	Writes fractions to represent probability

My Spin, I Win

Grade Level: Upper

MATERIALS: Three spinners for each group (see appendix a). Spinners should be divided into thirds

ORGANIZATION: The class should be divided into teams of three students.

PROCEDURE: Assign each team member a letter "A," "B," or "C," and provide a spinner for each student. Number the spinners as follows:

A = 8,6,1   B = 7,5,3   C = 9,4,2

Spin of A	Spin of B	A Wins	B Wins
8	7	x	
8	5	x	
8	3	x	
6	7		x
6	5	x	
6	3	x	
1	7		x
1	5		x
1	3		x

Have each student speculate as to which spinner will win most often.

Students A and B spin their spinners, and the winner is the person who spins the highest number. Repeat several times and record the findings.

Based upon the results, students should decide who will win between player A and C. Carry out the experiment to verify their conclusions.

Have students create tables similar to that given which demonstrate the probabilities of this contest. Students should then transfer this information from tabular form to a fractional form.

 Hands On, Inc
2121 Rebild Drive
Solvang, CA 93463

22	Constructs double bar graphs

The Long and Short of It

Grade Level: Upper

MATERIALS: Paper, pencil, graph paper, colored pens, class list

ORGANIZATION: Groups of four students

PROCEDURE: Gather data of heights of each student in the room. Record data on a class list so it is available to all groups.

Arrange the girls' and boys' heights in order -- from shortest to tallest. Create a bar graph which depicts this ordered size for boys and girls; however, alternate boys and girls on each graph line (as shown). It is helpful to use two different colors.

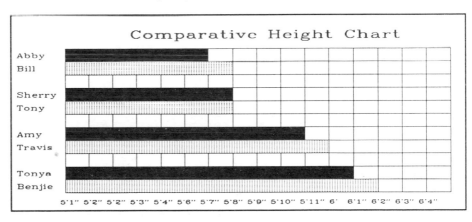

Comparative Height Chart

Discuss this graph with students as a beginning form of a double bar graph.

As a class project, have students create a large graph of this information and then create a human graph of students to compare and contrast.

Many discussion possibilities present themselves in this activity including a discussion of the average difference in height between boys and girls, difference in inches between boys and girls, and total number of inches for each gender and for the total class..

Hands On, Inc
2121 Rebild Drive
Solvang, CA 93463

22	Constructs double bar graphs

The Sports Connection

Grade Level: Upper

MATERIALS: Copies of a newspaper sports section for each student or each group of students.

ORGANIZATION: May be done individually, or in groups of two, three, or four.

PROCEDURE: Given a set of box scores or league leading statistics, students will create a double bar graph to represent information such as hits per times at bat, shots made for shots taken, completions per passes attempted, home runs in relation to hits, or other comparisons as selected by the students.

Each student should prepare a paragraph explaining the data presented in the double bar graph.

The teacher may wish to extend this assignment to create probabilities expressed as fractions. For example, a quarterback who had attempted 348 passes and had completed 170 might be said to have a 1 out of 2 or 1/2 probability of completing passes in his next game.

An important point to demonstrate to students is the similarity between percentage, probability, average, and fractions as concepts depicting similar types of data.

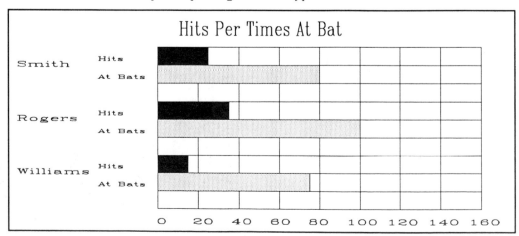

Hands On, Inc
2121 Rebild Drive
Solvang, CA 93463

22	Constructs double bar graphs

You Take My Breath Away

Grade Level: Upper

MATERIALS: Stopwatch or wrist watches with second hands, straws (two different types or colors), glue, scissors, rulers, and construction paper

ORGANIZATION: Teams of two, three or four students

PROCEDURE: Most students are familiar with the term "aerobics." This lesson uses aerobics as the basis for a lesson on double bar graphs.

Begin by dividing the class into groups and have each student count their number of heartbeats for one minute. They should record this number. Most students will range from 50 to 80 beats per minute.

Next, tell the students that they are going to do an aerobic workout for one minute. Explain that the purpose of aerobics is to increase the heartbeat, through exercise, in order to provide extra oxygen to the blood and to thereby exercise the heart.

Have students do one minutes of jogging in place, letting students select their own pace. At the end of three minutes, have them once again count the number of heartbeats per minute. Give them some time to rest and then repeat the exercise for five minutes, then seven, then nine, each time recording their heartrate after exercising.

From this information, they will be creating a three dimensional double bar graph using straws. This process is challenging because each student will have to create his own "scale." For example, each half inch (or centimeter) might be equivalent to 20 heartbeats. Once a straw segment is cut for each heart rate, have students glue this onto construction paper and label.

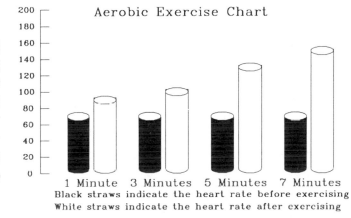

Aerobic Exercise Chart

Black straws indicate the heart rate before exercising
White straws indicate the heart rate after exercising

Hands On, Inc
2121 Rebild Drive
Solvang, CA 93463

22	Constructs double bar graphs

Fitness Counts

Grade Level: Upper

MATERIALS: Tally sheets, graph paper, stop watches (optional), measuring tapes

ORGANIZATION: Teams of two

PROCEDURE: This is a nice activity which fits well with a physical education program for student improvement.

An attempt should be made to remove the competitive part of PE from this activity. Tell students that they will only be measuring their own progress over a week long period. Have each student do activities such as broad jump, running race, ball throwing, pullups, and situps and record one another's results.

During each day's PE period, allow students to practice and become more efficient at these tasks. At the end of one week's time have students do the activities again and prepare a double bar graph which shows their individual improvement.

This might be continued as an ongoing project from one week to the next. Students will begin to see the value of learning to use graphs to chart progress.

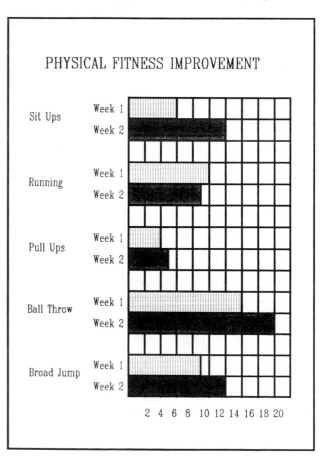

Hands On, Inc
2121 Rebild Drive
Solvang, CA 93463

23	Reads and interprets broken line graphs

Broken Line Graph Hunt

Grade Level: Upper

MATERIALS: Newspapers and magazines, especially USA Today, Time, Newsweek, or U.S. News and World Report

ORGANIZATION: Individually or in teams of two

PROCEDURE: Hand out magazines and have students cut out various types of broken line graphs. Glue these cutouts on separate sheets of paper and have students create questions about their graphs.

These student generated graph pages can be circulated throughout the room for response and practice by each individual. You may wish to laminate these papers for future use.

Sample Questions:

1. How can you tell which item in the graph is the most popular (most used, most important)?
2. How can you tell which item on the graph is the least popular?
3. How much longer is the longest bar than the shortest? How can you figure this out?
4. What is the value of the most popular (most used, most important) item on the graph?
5. Which item on the graph is used an "average" amount of time? How do you know?
6. What are the benefits of putting this information in graph form?
7. Do you find graphs to be a helpful way of presenting information? Why?

There are several task analysis items involving "reading and interpreting" various types of graphs. The intent of this book is to provide activity based math lessons and in general it is difficult to do reading and interpreting in an activity approach; therefore, we have provided universal lessons for these T.A. items and suggest you use these lessons in conjunction with your basic math series or with duplicated material.

Hands On, Inc
2121 Rebild Drive
Solvang, CA 93463

Weather Graphs

Grade Level: Upper

MATERIALS: Copies of five days of weather reports from a newspaper, graph paper

ORGANIZATION: Pairs of students

PROCEDURE: Hand out the information on weather for five days. Ask students to analyze the information and as a team try to decide on information which could be presented as a broken line graph.

As a class, discuss each team's selection as to why it will or will not work as a graph. Encourage students to try different approaches with different types of data. There are a variety of acceptable approaches.

Ask each team to plan its graph in terms of the value of the horizontal or vertical axis. Have each team draw its plan on the board. Discuss these plans with the entire class and then complete two or three of the graphs on the chalkboard or overhead.

From Table to Graph

Grade Level: Upper

MATERIALS: Table presented in appendix C, graph paper

ORGANIZATION: Can be done individually or in groups of two, three, or four

PROCEDURE: Hand out copies of appendix C to each student or group. Tell them that they will be constructing a broken line graph which depicts this information.

Essential elements to explain include: a title for the graph, labeling of the vertical and horizontal axis, neatness, and accurate transfer of data.

You may wish to work with the students in a step by step procedure or simply present the concept and have each group work together to arrive upon a finished product.

Hands On, Inc
2121 Rebild Drive
Solvang, CA 93463

24	Constructs a broken line graph

Weather or Not, You Want to Know

Grade Level: Upper

MATERIALS: Newspapers containing weather reports or almanacs with temperature, rainfall, and humidity information; graph paper and rulers.

ORGANIZATION: Can be done individually or in groups of two, three, or four.

PROCEDURE: Give each student or group a resource such as the newspaper or almanac and have them select some weather information which might be displayed as a broken line graph.

You may wish to combine this assignment with a science lesson on weather or a geography lesson on latitude and longtitude.

To make the assignment more of a learning tool, save weather sections for an extended period of time and let the students do comparative temperature graphs.

Upon completion of the graphs ask each student or each group to discuss why information is presented in graph form -- what makes "graphed" information more accessible?

You may also wish to make a differentiation between a broken line graph and a straight line graph. Please refer to the glossary for a description of these two graph types.

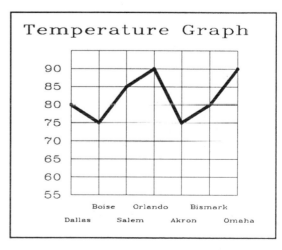

Hands On, Inc
2121 Rebild Drive
Solvang, CA 93463

24	Constructs a broken line graph

What a Line!
Grade Level: Upper

MATERIALS: Tape recorders with microphones, graph paper for wall displays

ORGANIZATION: Individually or in groups of two

PROCEDURE: This lesson will tap the students' creativity as they write and record a story problem which is depicted in a broken line graph. The teacher may want to duplicate the graph and story problem on this page as a guide for the students.

Explain that a broken line graph is really a story represented by a picture. The picture can shows the changes in a story from one given period of time to the next.

In this activity, students will need to do three things: first, they will need to generate a story line to work from; second, they will graph the day to day (week to week, etc.) changes which occur in the story; third, they will write and then record their story for the class.

When the project is complete, have students display their graphs and allow the entire class to listen to the taped story. This activity provides a very concrete experience in interpreting and understanding the motion of a broken line graph.

John, Robert, and Simon all started work on the same day, for the same number of hours, and for the same rate of pay. They each received $4.00 per hour for a 40 hour week. For the first week, everything went well, but during the second week, John's pay increased $.25 per hour and Robert's pay decreased $.50 per hour.

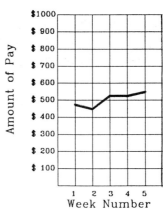

During the third week, Robert's pay increased to his original pay and Simon's pay went up $.50 per hour. In week four, everything remained the same, but in week five, all three men's pay increased $.25 per hour.

The broken line graph shows the increases and decreases from week to week.

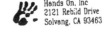

Hands On, Inc
2121 Rebild Drive
Solvang, CA 93463

24	Constructs a broken line graph

Working on Homework

Grade Level: Upper

MATERIALS: Graph paper and rulers

ORGANIZATION: Individually or in pairs

PROCEDURE: Explain to students that this activity will be an ongoing project in which they graph the percentage of homework assignments turned in each day by the entire class.

Have students select a point of intersection on the graph paper about one inch in from the left edge and one inch up from the bottom of the page. From this point have students count up towards the top of the page by "fives" -- labeling each box as they count. Along the horizontal (or x) axis, have them rule a line to within an inch of the right hand edge.

Each line will represent a day's homework assignments, or if there are multiple assignments, several lines can be used.

Each day, have students compute the percentage of assignments turned in by the whole class. After each entry, students should connect it with the line from the previous day.

Extensions include a homework report to parents or research as to the type of homework which students complete most successfully and least readily. There is a variety of such findings that give students a purpose for creating this graph.

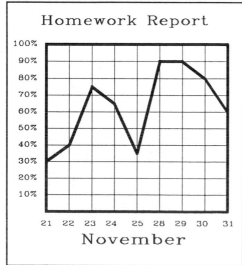

Hands On, Inc
2121 Rebild Drive
Solvang, CA 93463

24	Constructs a broken line graph

Taking Stock

Grade Level: Upper

MATERIALS: Financial page of a major paper (preferably a number of papers from the past five days) and graph paper.

ORGANIZATION: Best done in teams of two or three students.

PROCEDURE: Discuss the various types of information included in the financial page. Dow Jones Averages, stock results, precious metals prices, and mutual fund prices are all useful for this exercise.

Have each group select a specific stock or index to graph.

Hand out information for a period of five days and have students create a broken line graph which depicts this data. Have students predict the trends for the next few days.

This can be used as a very worthwhile ongoing project and can be integrated with economics and political science studies. You might also consider asking a local stockbroker to come in to talk with your class regarding the math involved in investing.

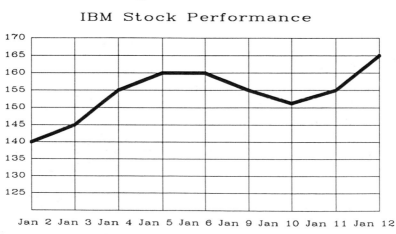

IBM Stock Performance

 Hands On, Inc
2121 Rebild Drive
Solvang, CA 93463

25	Reads and interprets circle graphs

Circle Graph Hunt

Grade Level: Upper

MATERIALS: Newspapers and magazines, especially USA Today, Time, Newsweek, or US News and World Report

ORGANIZATION: Individually or in teams of two

PROCEDURE: Hand out magazines and have students cut out various types of circle graphs. Glue these cutouts on separate sheets of paper and have students create questions about their graphs.

These student generated graph pages can be circulated throughout the room for response and practice by each individual. You may wish to laminate these papers for future use.

Sample Questions:

1. How can you tell which item in the graph is the most popular (most used, most important)?
2. How can you tell which item on the graph is the least popular?
3. How much larger is the largest section than the smallest? How can you figure this out?
4. What is the value of the most popular (most used, most important) item on the graph?
5. Which item on the graph is used an "average" amount of time? How do you know?
6. What are the benefits of putting this information in graph form?
7. Do you find graphs to be a helpful way of presenting information? Why?

There are several task analysis items involving "reading and interpreting" various types of graphs. The intent of this book is to provide activity based math lessons and in general it is difficult to do reading and interpreting in an activity approach; therefore, we have provided universal lessons for these T.A. items and suggest you use these lessons in conjunction with your basic math series or with duplicated material.

Hands On, Inc
2121 Rebild Drive
Solvang, CA 93463

Weather Graphs

Grade Level: Upper

MATERIALS: Copies of five days of weather reports from a newspaper, graph paper

ORGANIZATION: Pairs of students

PROCEDURE: Hand out the information on weather for five days. Ask students to analyze the information and as a team try to decide on information which could be presented as a circle graph.

As a class, discuss each team's selection as to why it will or will not work as a graph. Encourage students to try different approaches with different types of data. There are a variety of acceptable approaches.

Ask each team to plan its graph in terms of the value of the horizontal or vertical axis. Have each team draw its plan on the board. Discuss these plans with the entire class and then complete two or three of the graphs on the chalkboard or overhead.

From Table to Graph

Grade Level: Upper

MATERIALS: Table presented in appendix C, graph paper

ORGANIZATION: Can be done individually or in groups of two, three, or four

PROCEDURE: Hand out copies of appendix C to each student or group. Tell them that they will be constructing a circle graph which depicts this information.

Essential elements to explain include: a title for the graph, labeling of various sectors, neatness, and accurate transfer of data.

You may wish to work with the students in a step by step procedure or simply present the concept and have each group work together to arrive upon a finished product.

Hands On, Inc
2121 Rebild Drive
Solvang, CA 93463

26	Uses a tree diagram to find the total number of possible outcomes

Gym Dandy

Grade Level: Upper

MATERIALS: White, orange, blue, and red construction paper and scissors.

ORGANIZATION: May be done individually or in groups of two, three or four students.

PROCEDURE: Cut out materials to represent two shirts (orange and blue) and three pairs of shorts (white, orange, and blue).

Given the two shirts and three shorts colors, have students demonstrate the number of possible uniform combinations. Use cutouts to demonstrate this (there are six possible combinations).

Add a white shirt to the combination possibilities (9 combinations).

Add a red shirt and red shorts and ask how many combinations are possible. Have students construct tree diagrams to prove that they have included all possibilities.

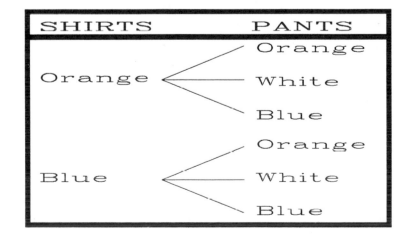

26	Uses a tree diagram to find total number of possible outcomes

Let it Roll

Grade Level: Upper

MATERIALS: Three dice, graph paper with 1 cm squares for each student or each group.

ORGANIZATION: Can be done individually or in teams of two.

PROCEDURE: Ask students to predict whether a "4" or "7" would occur most often if they were to roll two dice several times.

Have each student roll the dice thirty times and tally the result. Have each student or group of students create a tree diagram similar to that shown and ask them to explain why 4 and 7 should occur an equal number of times.

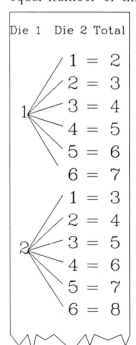

Which number, "4" or "7," would occur most often if you rolled the three dice? How can this experiment be presented as a fractional probablility?

Have students work from a factor tree similar to that presented. Get them started by providing the first column of information for two dice. Depending upon the level of the class, they may need varied amounts of guidance.

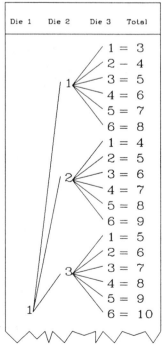

Hands On, Inc
2121 Rebild Drive
Solvang, CA 93463

26	Uses a tree diagram to find the total number of possible outcomes

A Handy Decision

Grade Level: Upper

MATERIALS: Graph Paper

ORGANIZATION: Divide students into groups of three

PROCEDURE: This activity is a decision making–probability game based upon the the rock/scissors/paper game format.

Have students imagine that the three of them are trying to select a gift to buy. There are three choices but they have money for only one. They need to decide which one they should purchase. Playing this game will allow them to make this choice.

Have each player make a fist and tap 1-2-3. At the count of "3," each student should display one, two, or three fingers.

The team then totals the number of fingers displayed. If the total is three or four, gift "A" is selected; if the total is 8 or 9, gift "C" is selected; if the total is five, six, or seven, gift "B" is selected.

Before beginning the game, each student should predict which outcome will occur most often.

Have students do the activity twenty to twenty-five times and tally the results. Compare the results to the predictions.

As the culminating activity, have students create a tree diagram to demonstrate the probabilities.

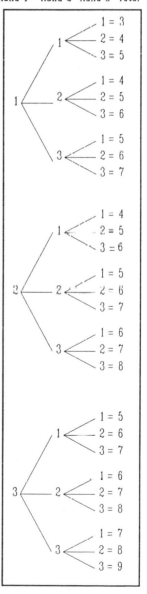

Hand 1   Hand 2   Hand 3   Total

Hands On, Inc
2121 Rebild Drive
Solvang, CA 93463

26	Uses a tree diagram to find total number of possible outcomes

Two For the Price of One

Grade Level: Upper

MATERIALS: Candies or colored tiles in a bag -- 2 red, 1 yellow, 1 green (for each group).

ORGANIZATION: Groups of two, three, or four

PROCEDURE: Show students the candies (tiles) as you place them in a bag in the front of the room. Ask them to answer the following questions before they begin pulling them from the bag:

1. What is the probability of getting two candies of the same color on one draw?
2. What is the probability of getting two candies of different colors on one draw?
3. What is the probability of getting a blue candy?

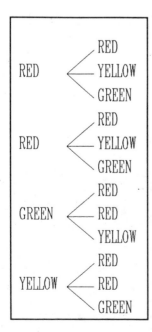

Have each group decide how they will tally results of each draw before they begin. The creation of a tally system is an important decision.

Have students pull two candies at a time for twenty trials. They should replace the tiles after each selection. Have them share their tally methods and their predictions/results on the chalkboard for group discussion.

Work through a tree diagram with the students to demonstrate a structured means through which to solve this problem.

COMBINATION PROBABILITIES			
RR	RY	RG	YG
2/12	4/12	4/12	2/12

 Hands On, Inc
2121 Rebild Drive
Solvang, CA 93463

27	Draws tree diagrams for a given problem

A Choice Car

Grade Level: Upper

MATERIALS: No special materials are necessary

ORGANIZATION: Whole class activity

PROCEDURE: Tell students that they will be purchasing a car called the Zephyr Zap and will need to select paint color, interior color and rally striping options. Their task will be to identify the number of possible combinations available.

Give them the following information: The Zephyr Zap comes in three colors -- red, white, or blue. The interior choices are silver, red, or black. Rally stripes are red or black but can only be done on cars with white paint.

With this information, ask students to construct a tree diagram which will identify all of the possible combinations.   Depending upon the experience of the class, you may have to help them get started with this assignment. The completed diagram is given although students may begin with exterior colors or interior colors. They should be encouraged to try different methods.

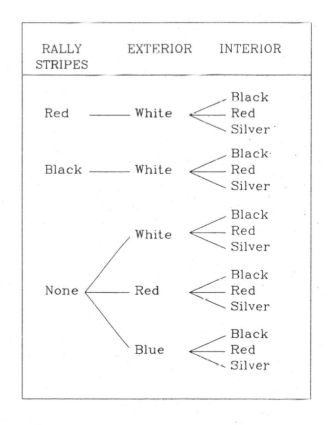

Hands On, Inc
2121 Rebild Drive
Solvang, CA 93463

27	Draws tree diagrams for a given problem

Fun...But Fair?

Grade Level: Upper

MATERIALS: Chips or coins marked with "A" on one side and "B" on the other.

ORGANIZATION: Students should be divided into teams of two.

PROCEDURE: Each student should be given a chip. Each player flips a chip at the same time. Player 1 gets a point if the chips match. Player 2 gets a point if the chips do not match.

```
CHIP 1  CHIP 2  WHO WINS

        A ──── 1
   A <
        B ──── 2

        A ──── 2
   B <
        B ──── 1
```

Have the students play a game to ten and decide if the game is fair or unfair. Have them create a graph, table or chart to demonstrate their decisions as to the fairness of the game.

Students will need help in getting started on the charting of this information, and you may wish to do this as a "whole class" activity. Once students have worked with displaying the information, present the tree diagram as shown.

Do this same activity with three chips -- one marked "A" and "B;" one marked "A" and "C;" and one marked "B" and "C."

Flip all three chips at once. Player 1 gets one point if there is a match of two chips; player 2 gets a point if there is no match.

Have students decide if this game is fair or unfair and create a graph, table or chart to prove these conclusions.

```
CHIP 1   CHIP 2   CHIP 3    WHO WINS

                    B ──── 1
              A <
                    C ──── 1
        A <
                    B ──── 2
              C <
                    C ──── 1

                    B ──── 1
              A <
                    C ──── 2
        B <
                    B ──── 1
              C <
                    C ──── 1
```

Hands On, Inc
2121 Rebild Drive
Solvang, CA 93463

27	Draws tree diagrams for a given problem

A Horse of a Different Color

Grade Level: Upper

MATERIALS: Three different colors of construction paper cut into small squares. Provide ten to fifteen squares of each color for each student.

ORGANIZATION: This activity can be done by students individually or in groups of two, three, or four.

PROCEDURE: Students will be "shoeing" an imaginary fashion conscious horse who wears a different color shoe on each of three feet and no shoe on the fourth.

Using the construction paper, students will need to decide how many fashion outfits can be designed for the horse. A different outfit is counted in any possible combination of colors as long as all three colors are used and there is one non-shod foot.

Since students do not have a sufficient number of colored squares to display the entire array of 72 combinations they will need to decide upon a method to solve the proposed problem.

Regardless of the method students use to prove their findings (graphing, tables, or arrays) they should ultimately come to the conclusion that a tree diagram in combination with use of multiplication is probably the most effective method of solving such a probability problem. You should direct them to this discovery.

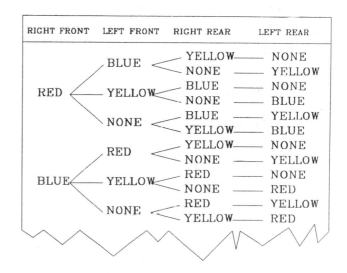

Hands On, Inc
2121 Rebild Drive
Solvang, CA 93463

27	Draws tree diagrams for a given problem

A Choice Taco!

Grade Level: Upper

MATERIALS: Construction paper of various colors and shapes such as 7" circles, 6" semicircles, and 2" by 3" rectangles -- these represent various taco parts

ORGANIZATION: Small groups of two or three students

PROCEDURE: Have students pretend that they are working at the local taco shop creating new tacos for the customers. Using the "rules" list below, have students create the various taco combinations and then do a tree diagram which graphically represents these probabilities.

Tortilla types   -- corn and flour (yellow and white round construction paper)
Fillings   -- beef, chicken, and beans (brown, black, and red construction paper semicircles)
Toppings   -- cheese, tomatoes, lettuce, hot sauce, guacamole (yellow, red, green, orange, and black rectangles)

Basic rules include the following: You must use one tortilla, one filling, and two toppings on each creation.

As an extension you might have students find the various possibilities if you do not limit the number of toppings. Some very interesting tree diagrams may be created.

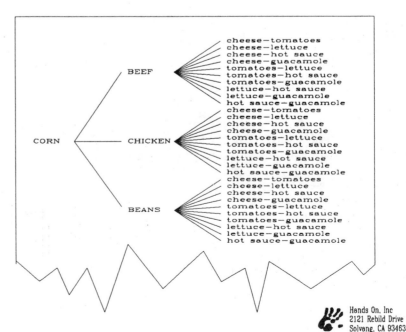

Hands On, Inc
2121 Rebild Drive
Solvang, CA 93463

28	Predicts results of simple probability experiments by multiplying the probability of favorable outcomes by the total number of outcomes

Getting Your Money's Worth

Grade Level: Upper

MATERIALS: Five phony dollar bills (denominations of a one, five, ten, twenty, and fifty), a sack, and scratch paper for each group.

ORGANIZATION: Teams of two, three or four students

PROCEDURE: Place a set of five bills in each sack and distribute to each group. Ask the class the probability of picking out the fifty dollar bill. Their response should be "one in five."

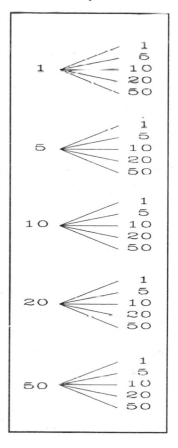

If you replace the first pick, what is the probability of drawing a fifty dollar bill on the second draw? Once again the chance is one in five.

Ask students the probability of drawing the fifty on two consecutive draws. Is there a way that students can solve this problem? You may wish to construct the tree diagram as given or allow students to create their own methods of answering these questions.

As a final point, demonstrate to students that the most efficient method of solving this problem is to multiply the fractions (1/5 x 1/5).

$$\frac{1}{5} \times \frac{1}{5} = \frac{1}{25}$$

For advanced students you might ask them to predict the probabilities if bills are not returned to the bag.

Hands On, Inc
2121 Rebild Drive
Solvang, CA 93463

28	Predicts results of simple probability experiments by multiplying the probability of favorable outcomes by the total number of outcomes

This Will Make You Flip

Grade Level: Upper

MATERIALS: A coin and a tally sheet for each student; scratch paper will also be needed.

ORGANIZATION: Can be done individually or in teams of two, three, or four

PROCEDURE: Before beginning this lesson you will need to review the procedure for multiplying fractions.

Provide a coin for each student and ask, "What is the probability that heads will occur on the first toss?" The response should be 1 in 2 or 1/2.

Then ask students to predict the probability of heads occuring on the second toss. The chances again are one in two (or 1/2). What is the probability of heads occuring on the first two tosses? Guide students to realize that there are four possible outcomes in two tosses -- heads/heads, heads/tails, tails/heads, and tails/tails. Therefore the chances are 1 in 4 or 1/4. Can students figure out how this might be completed mathematically?

$$\frac{1}{2} \times \frac{1}{2} = \frac{1}{4}$$

Demonstrate that since probability can be represented as a fraction, it can also be manipulated as a fraction. This being the case, what is the probability that heads will occur on three consecutive tosses?

$$\frac{1}{2} \times \frac{1}{2} \times \frac{1}{2} = \frac{1}{8}$$

As a final step, ask students to describe the number of "flips" in this example. Discuss probability of 1 in 256 and have teams of students conduct the experiment and see if anyone gets 8 "heads" in a row.

$$\frac{1}{2} \times \frac{1}{2} \times \frac{1}{2} \times \frac{1}{2} \times \frac{1}{2} \times \frac{1}{2} \times \frac{1}{2} \times \frac{1}{2} = \frac{1}{256}$$

Hands On, Inc
2121 Rebild Drive
Solvang, CA 93463

28	Predicts results of simple probability experiments by multiplying the probability of favorable outcomes by the total number of outcomes

Out For a Spin

Grade Level: Upper

MATERIALS: Spinners sectioned into thirds (blue, red, green) and tally sheets for each team

ORGANIZATION: Teams of two

PROCEDURE: Distribute spinners to each group and have students determine the probability of each result (1 in 3). Have them spin three times and tally the results. On the chalkboard, record the results of each group. Results will vary greatly.

Ask students to decide how many different possible combinations there are for three consecutive spins. They might choose to create a table or a tree diagram, but the most effective way would be to multiply 1/3 x 1/3 x 1/3 (there are 27 possible combinations).

	Group 1	Group 2	Group 3	Group 4	Group 5
Spin 1	Red	Green	Green	Blue	Red
Spin 2	Red	Red	Blue	Green	Blue
Spin 3	Green	Blue	Blue	Red	Green

Explain that just as 1/2 x 1/2 solved the problem in the previous lesson, manipulation of fractions makes the process less time consuming. Many students will need to verify the figure of 27 by writing out the possibilities. This activity is encouraged.

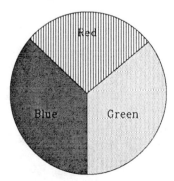

For students who are comfortable with the outcome, pose some extensions -- what if there were 4 colors on the spinner, what if the spinner colors were in 25%, 25% and 50% segments.

Hands On, Inc
2121 Rebild Drive
Solvang, CA 93463

28	Predicts results of simple probability experiments by multiplying the probability of favorable outcomes by the total number of outcomes

A Chance to Save a Life

Grade Level: Upper

MATERIALS: A roll of lifesavers (or other covered/assorted candies) for each student or each group

ORGANIZATION: Can be done individually or in teams of two, three or four students

PROCEDURE: Hand out the lifesavers and tell students that they are going to predict the color life saver on each end of the roll. Discuss the various things which students must know before making a guess -- the flavors possible and perhaps the number in the pack.

The purpose of this lesson is to do research so the next time students open a roll, they will be able to make an educated guess -- based upon the probabilities they discover.

Let each group guess and then look at each end. Before removing (or eating) the lifesaver, use the information they have gathered to guess the color of the next lifesaver. Continue this procedure moving through the pack. Students should keep the lifesavers in the order in which they are removed.

Have students record the information and, as a class, chart the color patterns in the roll. A wide variety of information can be gathered but basically students should use the five flavors, repeats of color patterns, and the number of candies in a roll to create various probabilities.

A wide variety of extensions are available including tree diagrams, zero probability, and average/range/median/mode lessons.

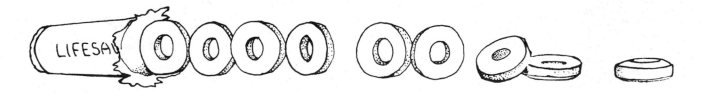

Hands On, Inc
2121 Rebild Drive
Solvang, CA 93463

29	Identifies zero probability as an event which cannot occur and a probability of one as an event which is certain to take place

An Unfair Fair

Grade Level: Upper

MATERIALS: A jar with an opening a bit too small for a dime, penny, or other coin, paper clips, paper, a bucket or large beaker, a smaller beaker, saran wrap, red fish cutouts, and a magnet

ORGANIZATION: A whole class activity -- later break into small groups

PROCEDURE: The purpose of this activity is to demonstrate to students that probability exists in an event which always occurs (probability of one) and an event which cannot occur (probability of zero).

To demonstrate this, we have created four carnival games. You may demonstrate these for the whole class and have students come forward to "compete" for prizes.

GAME 1 - DIME TOSS
Place a bottle -- with an opening just small enough so that a coin will not drop in -- in the front of the room. Have students stand above the bottle and try to drop coins into the container. Don't allow them to measure. If you want to give more credibility to the game, you might file one edge of a coin and drop it in the bottle beforehand. Have several students try to complete the task; however, they will never drop the coin in.

GAME 2 - PAPER CLIP DROP
Place a sheet of paper in the front of the room and have students hold a paper clip one to two feet above the paper. The object is to drop the clip and have it land on the paper. They should accomplish this feat 100% of the time.

Hands On, Inc
2121 Rebild Drive
Solvang, CA 93463

GAME 3 - FISHING FOR COLOR

Cut out five to ten red fish and place a paper clip on each. Block the students' vision so that they cannot see that all of the fish are red. Create a fishing pole by tying a string to a magnet and to a stick. Tell students that the object of the game is to catch a red fish -- any other color does not win a prize. You might act surprised or disappointed as students continually win by catching red fish. You then reveal that all the fish are red.

GAME 4 - COIN IN THE BUCKET

Fill a small and large beaker with water. Place saran wrap over the small beaker and place this small beaker on the bottom of the large beaker. The object is to have students drop a coin in the large beaker and have it drift into the small beaker. Obviously, the saran wrap will not allow this to happen.

When you have finished with the experiments ask students to discuss the carnival games you have played. You may have to explain why the games always were won or were never won. Students might want to come up with variations which would provide probabilities other than zero and one.

An extension to this activity would be to ask students to prepare carnival booths for a NO WIN - NO LOSE carnival in the room. As teams, they can create booths which appear to offer challenges but in reality cannot be won or lost. The purpose being to demonstrate that the probability of zero and the probability of one exist and need to be understood and evaluated.

 Hands On, Inc
2121 Rebild Drive
Solvang. CA 93463

29	Identifies zero probability as an event which cannot occur and a probability of one as an event which is certain to take place

Pounds and Years Bring No Cheers

Grade Level: Upper

MATERIALS: A bathroom scale and graph paper

ORGANIZATION: A whole class activity

PROCEDURE: Have ten students volunteer to be weighed and record their weights on the chalkboard or overhead. Next to each student's weight, have them record their ages in months.

Each student should then arrange the weights from least to greatest and create a double bar graph representing this information.

Given the information on the double bar graph, have students postulate on the probability that age is indicative of weight. The obvious response is that there is no correlation between weight and age (after a certain age). Identify that the zero probability or zero correlation demonstrated in this graph is useful in that it still proves a hypothesis.

This lesson can be extended to include discussion of average, range, median, and mode.

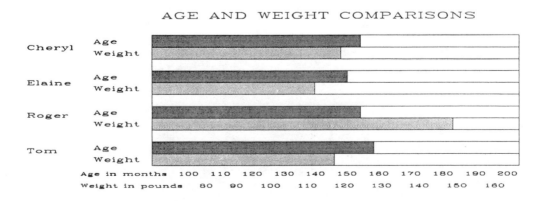

29	Identifies zero probability as an event which cannot occur and a probability of one as an event which is certain to take place

A Remarkable Feet!

Grade Level: Upper

MATERIALS: Cut out 10 blue and 10 black socks for each group.

ORGANIZATION: Teams of four

PROCEDURE: Create a scenario of a businessman who awakens every morning and selects either black or blue socks to wear with his suit. Because he is an organized businessman, he has two sock drawers. In DRAWER 1 he places all of this black socks, and in DRAWER 2 he places his blue socks. Have the students separate their socks to duplicate this organization.

Ask each group to identify the probability of selecting a black sock from drawer 1 (probability of 1); the probability of selecting a blue sock from drawer 2 (probability of 1); the probability of selecting a black sock from drawer 2 or a blue sock from drawer 1 (zero probability).

From this point forward have students experiment with various assortments of black and blue sock drawer distribution. Tree diagrams might be an effective way to demonstrate this.

As a final point, have students identify that the only way to assure a probability of one is to have no other options available; the only way to have a zero probability is if none of the anticipated outcomes is possible.

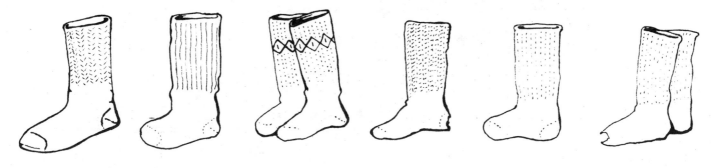

Hands On, Inc
2121 Rebild Drive
Solvang, CA 93463

30	Identifies whether to use a sample or census in a given situation

A Calculated Guess

Grade Level: Upper

MATERIALS: A large jar of beans, red hots, or other small, difficult to count items in a container; rulers, or other means of measurement

ORGANIZATION: Teams of four

PROCEDURE: Show a sample container of countable items and ask students to decide how they might **reasonably** estimate the total number of items in the jar. Obvious responses include counting, making a guess, or (hopefully) taking a sample.

Discuss various means of doing a sample which might include measuring fractional parts of the jar and then counting only one of these parts, or spreading the entire contents over the desk and then dividing the area covered into fractional parts. Students may come up with other ideas as well.

Students should be able to identify that a guess, a reasonable estimate, a sample, and a census (counting each item) are possible methods to arrive upon a total number of contents and that each method becomes increasingly more accurate.

Have students generate a variety of situations in which these various approaches might be useful.

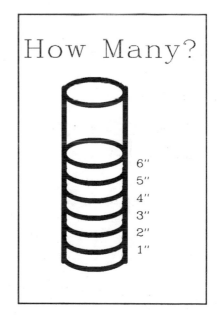

Hands On, Inc
2121 Rebild Drive
Solvang, CA 93463

30	Identifies whether to use a sample or census in a given situation

My Kind of Song

Grade Level: Upper

MATERIALS: No special materials are needed for this activity

ORGANIZATION: To be done as individuals or in groups of two, three, or four.

PROCEDURE: Begin this lesson by describing the difference between a census and a sample. To clarify, a census is a survey of all people; a sample is a survey of a representative percentage of people.

Students will be gathering information on favorite music styles and groups within the classroom. It is best if choices are limited to three possibilities.

This information should be collated and organized into Venn diagrams.

Based upon the results of the Venn diagram, students should predict probable ratios and probabilities of this same survey in other classes.

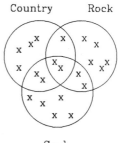

Identify that, in this case, students have performed an in-class census, but this represents only a sample of the entire population. This is an important concept which students need to understand.

You may wish to poll other classes to prove the accuracy (inaccuracy) of a sample.

21 TOTAL STUDENTS	
Rock	5
Soul	5
Country	4
Rock/Country	1
Rock/Soul	2
Country/Soul	2
Rock/Country/Soul	2

Hands On, Inc
2121 Rebild Drive
Solvang, CA 93463

30	Identifies whether to use a sample or census in a given situation

Take That, You Drip!

Grade Level: Upper

MATERIALS: Cups, glasses (or other containers), eyedroppers, and water

ORGANIZATION: Individual or in teams of two or three students

PROCEDURE: Review the definition of census and sample -- a census being a total count and a sample a count of a representative portion.

Hold up a glass of water and ask the students to guess how many drops of water are contained in it. Discuss how they might reasonably find the answer. Responses might include guessing, using a dropper and counting all the drops, or counting the drops for a sample portion of the water in the cup.

Have each group or individual test his idea with the dropper to arrive upon the most efficient method. You may wish to make a differentiation between speed and accuracy. A sample provides a faster solution, but a census is more accurate.

As an extension you may wish to discuss the ten year census carried out by the federal government and discuss why it is necessary to do a census occasionally. You may organize the class to do a census of the school or to perhaps make estimates of the ages of students based upon the sample within the classroom.

How Many Drops Must Drip

Hands On, Inc
2121 Rebild Drive
Solvang, CA 93463

30	Identifies whether to use a sample or census in a given situation

Mingling With the Bugs

Grade Level: Upper

MATERIALS: An area where students can count blades of grass in a given area, tape measure, other measuring devices.

ORGANIZATION: Teams of four

PROCEDURE: Begin the lesson by discussing the difference between a census and a sample. A census is a poll of all participants, a sample is a poll of a representative portion of a group. The discussion needs to include reference to when a census is necessary and when a sample will suffice.

Tell students that they are going to count the blades of grass in a particular area of the school. Obviously, a census -- counting each blade is impractical -- therefore students should develop some strategies for developing a sample.

They should be guided to measure the overall area and divide it into grids. Each group can take a certain grid to actually count blades of grass. Students will select a variety of sizes to measure and this should be part of the problem solving activity.

Upon completion of the estimates, re-focus discussion on the need for precision in some activities and the need for speed of data collection in others. The point of this lesson is to heighten awareness of the benefits of a sample vs. a census.

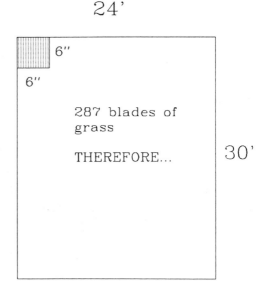

School Quad

Hands On, Inc
2121 Rebild Drive
Solvang, CA 93463

31	Constructs a circle graph given data

Family Finance -- Budget

Grade Level: Upper

MATERIALS: Family Finance cards (see Appendix B), list of budget items, rulers, and compasses

ORGANIZATION: Please see Appendix B for information on setting up this activity.
This is lesson 1 on the Family Finance unit.

PROCEDURE: Look at the family income and the list of items to be budgeted. Write the dollar amount to be budgeted and compute the percent of the total beside each item. You may wish to provide newspapers, want ads, and advertising to give students a sense of cost for each budgeted item.

Construct a circle graph that shows each percent and label accordingly.

Rules to consider for budgeting:
1. Allow at least 1/3 but not more than 1/2 of your budget for housing.
2. Allow at least 1/4 but not more than 1/3 of your budget for food.
3. Allow at least 1/10 of your income for utilities.
4. Put enough money into savings because holiday spending and a vacation will come from the savings account.
5. Each item must be budgeted

Have students create a paragraph which describes the method they used to create their budgets.

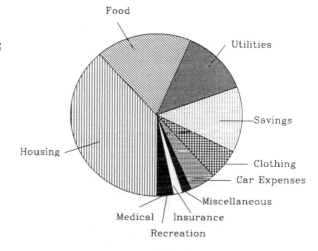

YEARLY BUDGET WITH INCOME OF $20000			
ITEMS	YEARLY	MONTHLY	%AGE
Housing	7000	583	35%
Utilities	2000	167	10%
Medical	500	42	2.5%
Food	5000	417	25%
Recreation	500	42	2.5%
Insurance	500	42	2.5%
Clothing	1000	83	5%
Savings	2000	167	10%
Car expenses	1000	83	5%
Miscellaneous	500	42	2.5%

 Hands On, Inc
2121 Rebild Drive
Solvang, CA 93463

31	Constructs a circle graph given data

Family Finance -- Vacation

Grade Level: Upper

MATERIALS: Family Finance lesson as explained in Appendix B. Books and brochures listing costs of motels, meals, entrance fees, compasses

ORGANIZATION: Groups of four

PROCEDURE: Tell students to imagine that they have been saving for a vacation for the past 10 months. They may use 1/2 of the savings to plan their family vacation. List the cost of motels, food, admissions, fees, transportation and other expenditures.

Based upon this information, have students compute the percentage of total expenditure for each item and make a circle graph to depict these expenditures.

Upon finishing the graph, students should be asked to reflect upon the type of vacation and the expenditures for such an excursion. There are many discussion items which can be addressed in this type of activity.

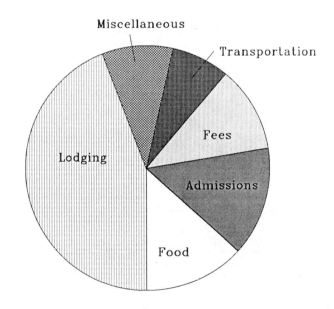

Hands On, Inc
2121 Rebild Drive
Solvang, CA 93463

31	Constructs a circle graph given data

Family Finance -- Holidays

Grade Level: Upper

MATERIALS: Family folder, catalogues for selecting holiday gifts, and compasses

ORGANIZATION: To be done individually with each student's fictitious family (see Appendix B)

PROCEDURE: Look at the budgeted savings left after your vacation trip. Use 3/4 of the funds left for Christmas presents. Determine and write what part of that amount you will give the children to spend. Subtract that amount. You must give presents to the children, your spouse, relatives such as grandparents, and friends.

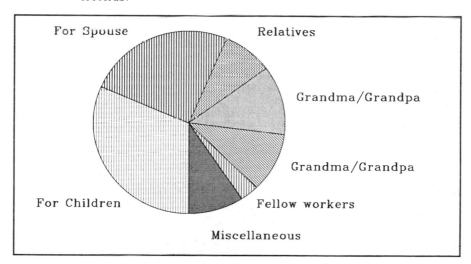

Make a list of people to receive presents and what their gifts will be. You cannot spend more than you have. Use catalogues to help you determine costs. Make a circle graph showing what percent of the total amount is spent on each group.

Upon completion of the graph, write a paragraph discussing the difficulties encountered and how these problems might be solved.

 Hands On, Inc
2121 Rebild Drive
Solvang, CA 93463

31	Constructs a circle graph given data

Family Finance –– Clothing

Grade Level: Upper

MATERIALS: Family folder, catalogues for selecting clothing, compasses

ORGANIZATION: To be done individually with each student's fictitious family (see Appendix B)

PROCEDURE: Look at the amount budgeted for clothing. Your assignment is to buy summer clothing for each member of your family. List each family member and the items of apparel he/she will need.

Using catalogues to determine prices, use two months of the clothing budget for this task. Compute the total amount spent on each person and construct a circle graph to that show the percentage spent on each individual.

Discussion of the graphs might include: Is there a difference in the percentage spent on each family member? Why is this true? Was the budgeted amount sufficient? If not, what could you change to allow for more money for clothing?

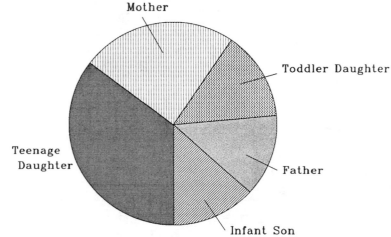

Hands On, Inc
2121 Rebild Drive
Solvang, CA 93463

32	Makes predictions based on line of best fit

Swingin' String

Grade Level: Upper

MATERIALS: Washers, string, rulers, measuring tapes, masking tape, watches with second hands, recording sheets

ORGANIZATION: Whole class activity or in teams of four

PROCEDURE: The teacher should begin the lesson by demonstrating the construction of a pendulum (as shown). Ask students to predict the number of swings the washer will make in 10 seconds.

To standardize the process on the first trial have all students measure the string at 12 inches, there is one washer, and the washer is dropped from a position even with the ruler (a 3:00 o'clock position). Students should record all data. Have groups vary different aspects of the experiment.

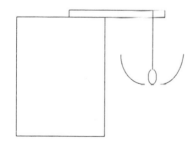

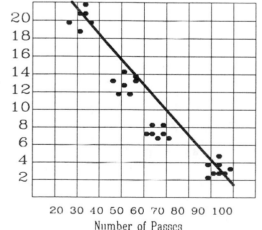

One group might shorten the string each trial while a second group lengthens the string; one group might add a washer each trial; and one group might drop the washer from a different height each trial.

Have students construct a line of best fit graph similar to that shown. Given this information have students make predictions as to the number of swings for various string lengths and various washer weights.

Hands On, Inc
2121 Rebild Drive
Solvang, CA 93463

32	Makes predictions based on line of best fit

On the Rebound

Grade Level: Upper

MATERIALS: Any of the following materials will work well: soccer ball, ping pong ball, tennis ball, volleyball, golf ball, basketball; metric or yard stick and masking tape.

ORGANIZATION: This activity can be done in teams of two students or by the class as a whole.

PROCEDURE: Attach the meter stick to a table so that the 1 cm mark points toward ground level.

Hold one of the balls at the 100 cm mark and drop it. Have a student mark the high point of the first bounce. Record the results and bounce each ball several times to gain an accurate sample.

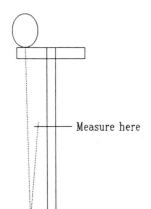

Measure here

Your students will find that not all bounces go to the same height. Draw a line of best fit keeping as many plotted points above the line as below the line

Have students predict the results of a ball dropped from 200 cm and from 50 cm. Prove and graph this information through the line of best fit..

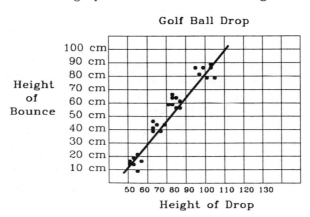

Hands On, Inc
2121 Rebild Drive
Solvang, CA 93463

32	Makes predictions based on line of best fit

What Measure Meant!

Grade Level: Upper

MATERIALS: Four or five containers for which you know the volume (a set of measuring cups is ideal), bags of beans.

ORGANIZATION: Groups of two, three, or four

PROCEDURE: Students will need to know how to plot coordinates. Have each student fill one of the containers and then count the number of beans that fit.

After each vessel has been filled and the beans counted, the students should create a coordinate graph and plot their results.

After plotting the points, the information should be transferred to a large piece of graph paper (or a graph on an overhead) so that a composite of the entire class is created. A line should then be drawn which intersects the center of each cluster of marks on the graph. This is the line of best fit.

Students should be asked to estimate the different information using the line of best fit. For example, they can estimate what size (volume) container it would take to hold "N" number of beans or they might estimate the number of beans that a given size container might hold.

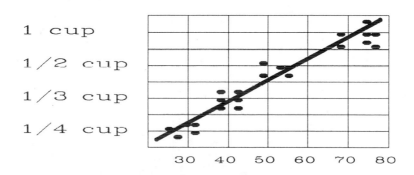

Hands On, Inc
2121 Rebild Drive
Solvang, CA 93463

32	Makes predictions based on line of best fit

Rollin' Rollin' Rollin'

Grade Level: Upper

MATERIALS: Four or five cylindrical items that roll such as pencils, pens, candy cane; stiff paper or cardboard, some books, a measuring tape for each group

ORGANIZATION: Groups of two, three, four, or as a whole class activity

PROCEDURE: Students will need to know how to create a coordinate graph so you may need to prepare for this lesson ahead of time.

To begin the lesson, students create an incline plane by propping one end of the cardboard up with books; it should not be too steep. They then take turns rolling items down the plane and measuring the distance travelled. After the information has been obtained, they should sort the items rolled and rank them by weight. Based upon the weight and distance travelled, create coordinate graphs to represent the results.

To create a line of best fit, they draw a straight line that stretches across the graph and intersects the cluster of points at the left side of the graph and the cluster at the right side of the graph.

This line can then be used to form estimations of relationships based on the data gathered. For example, by cross referencing the two points on the line, students might be able to estimate how far a certain object might roll depending upon its weight.

An extension might be to vary the height (slope) of the incline plane to see what effect it has on the experiment.

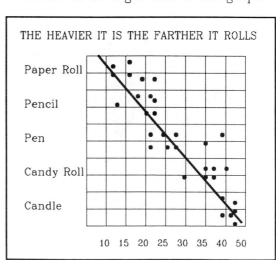

Hands On, Inc
2121 Rebild Drive
Solvang, CA 93463

33	Completes and constructs frequency tables

Three Coins in a Fountain

Grade Level: Upper

MATERIALS: Two bowls or shallow pans for each group -- one filled with water, one empty, a water pitcher, coins or markers to toss, paper towels for clean up

ORGANIZATION: Teams of three or four students

PROCEDURE: Tell students that they are going to perform an experiment which will be amazing to all who view it. They will be tossing coins into a bowl to see how many they can make stay vs. bouncing out. You may wish to experiment with the type of bowl you have supplied – you may wish to drop the coin from above or toss it from the side.

Place one empty bowl at each desk and have each group member try five times to toss the coin into the bowl and have it stay. Record results.

Ask them if they can think of any way to be more successful in this activity. You might let them try different approaches or different sized coins. Move about the room with a pitcher of water and fill each bowl. Have the groups repeat the experiment and record the results. Students will find that the coin stays in the bowl on every drop or toss.

Empty Bowl	Attempts	Filled Bowls
0	5	4
1	5	5
0	5	5
0	5	4
1	5	5

Given the results of the first trials (empty bowl) and the second trial (water filled bowl) have them graph or diagram this information in some way. A variety of responses will be presented by students. Allow them to be creative in their responses.

Show students the frequency table as pictured. Explain to students that such a chart reflects the probability of given outcomes.

You might have students write a paragraph explaining the difference between their chart and the frequency table which you described.

Hands On, Inc
2121 Rebild Drive
Solvang, CA 93463

33	Completes and constructs frequency tables

Side Splitting

Grade Level: Upper

MATERIALS: Sets of various polyhedral dice ranging from six sided to ?

ORGANIZATION: Groups of four

PROCEDURE: Each group of students is given a set of polyhedral dice and is asked to predict the probability (odds) of the various combinations of numbers. Six sided dice will give combinations ranging from 2 through 12. Have them estimate the number of times each number will occur in fifty rolls.

Have students tally while the rolling process occurs.

When they have finished, ask each group to discuss the accuracy and inaccuracy in their guesses. Ask them if they feel the number of rolls was sufficient for basing probabilities. The group with the six sided-dice will find that the sample is relatively accurate while the group with ten or twelve sided dice will find that their proof is not so evident.

Have students complete a frequency table which reflects the possible combinations of their dice as well as the frequency of each total (as shown). Discuss the differences between a sample and a frequency table and the usefulness of each approach.

TOTAL POINTS	POSSIBLE COMBINATIONS
2	1
3	2
4	3
5	4
6	5
7	6
8	5
9	4
10	3
11	2
12	1

Hands On, Inc
2121 Rebild Drive
Solvang, CA 93463

33	Completes and constructs frequency tables

Car Care

Grade Level: Upper

MATERIALS: Markers or crayons, white cutouts of cars

ORGANIZATION: Pairs or individually

PROCEDURE: Have your class imagine that they work in an auto body paint shop and are managers in charge of buying paint for the cars. You are curious as to how many combinations you can create given a certain number of colors.

For example, if you have two colors, you can do three different paint jobs -- color 1, color 2, and a combination of colors 1 and 2. Have students do this exercise picking any two colors.

Ask if there is any way for them to record this on a table or chart (a sample is given).

Next, have them see how many combinations they can create with three colors -- they may do this by coloring cars or in any other means they wish to use. Have them add this information to the chart.

Then have them do four and five different colors -- all the time adding the information to the table. See if they can find a pattern in this information. This may be difficult for them to see since the pattern is two times the previous result plus one.

Colors	Combinations
2	3
3	7
4	15
5	31
6	63

$$(3 \times 2 + 1 = 7 \qquad 7 \times 2 + 1 = 15 \qquad 15 \times 2 + 1 = 31)$$

The ultimate goal is for students to realize that a frequency table can help to provide a means by which to solve problems. It places things in an orderly structure which can be analyzed.

Hands On, Inc
2121 Rebild Drive
Solvang, CA 93463

33	Completes and constructs frequency tables

Hooping It Up!
Grade Level: Upper

MATERIALS: Basketball court and several basketballs

ORGANIZATION: Students working in teams of two to record results.

PROCEDURE: Shooting a basketball provides the data for the creation of a frequency table in this activity. Students will be shooting a one foot (from under the hoop), a five foot, and a fifteen foot (from the free throw line).

Begin in the classroom by dividing the class into teams of two students. They should make recording sheets as shown.

When the students go outside to shoot, each player should shoot ten shots from each distance listed above. The partner should record the number of made baskets for each distance.

When students return to the room, they should devise a way to represent this information as a table (frequency chart). There are several approaches that can be used (vertical, horizontal, multiple columns, individual tables, etc.).

Once all students have finished their tables, display them on a bulletin board and discuss various aspects of the data. For example: How might students improve upon the frequency of made baskets? How could they figure a "class average" for each distance? What generalizations can they make by looking at the data as a whole?

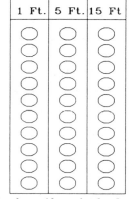

Darken the circle for each shot made

Hands On, Inc
2121 Rebild Drive
Solvang, CA 93463

| 34 | Uses multiplication to find total number of outcomes |

Did You Get the License Number?

Grade Level: Upper

MATERIALS: License plates from various states

ORGANIZATION: Individually or in groups

PROCEDURE: There are several travel games that students can play using license plates but they have probably never thought about how many combinations of letters and numbers are possible for each state.

Ask three or four students who know their parents' license plate numbers to volunteer the information. Write these plate numbers on the board. Their task will be to select one of the choices, and attempt to discover the maximum number of license plates which could be made using the letter and number configuration. This can even be done with the "customized" license plates offered by many states.

Before giving students any clues, let them brainstorm in small groups to find a way to solve this problem. The basic formula for any plate is to give a value of 26 to each letter and a value of 10 to each number (even though 1000 could not be used in a three digit plate, the number 000 could be used; but let the students discover this).

Once students have created a method for finding the number of possible solutions, have them write a paragraph explaining the process they used and why it works. Discuss the reason for using letters rather than all numbers.

As an extension have students design their own license plate that would allow 150,000,000 cars to be registered without any duplication. See which student can do this using the fewest number of license plate characters.

92   CALIFORNIA   Dec

2KWP930

| 34 | Uses multiplication to find total number of outcomes |

The Ski's the Limit

Grade Level: Upper

MATERIALS: Various colors of construction paper cut out to represent three hats, one coat, two scarves, and two pairs of pants

ORGANIZATION: Teams of two, three or four students

PROCEDURE: Tell students that they are planning a ski trip and since they want to be very fashionable on the slopes, they need to plan the various "outfits" which they can create.

Given the cutouts, allow each team to glue the various outfit combinations on a piece of construction paper. They must wear a hat, pants, jacket, and scarf with each outfit.

If the previous exercises have been done with the class, they should be able to figure out that 3 (hats) x 1 (coat) x 2 (scarves) x 2 (pants) will equal 12 different outfits.

For students who are more advanced or like a challenge, add various other types of gear to their ski lists. A second extension might include a lesson in color coordination related to an art lesson.

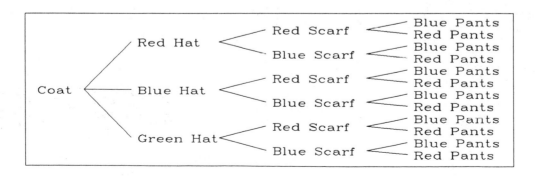

Hands On, Inc
2121 Rebild Drive
Solvang, CA 93463

34	Uses multiplication to find total number of outcomes

Crime Stoppers

Grade Level: Upper

MATERIALS: Pattern of a human face (similar to the one shown) duplicated on ditto paper

ORGANIZATION: Student teams of two or three students

PROCEDURE: Have students imagine that they are artists for the police department and they are to make a drawing of a known bank robber. Certain facts about his appearance are already known.

He has black hair and a pronounced widow's peak. He has a small scar on his left cheek but informants are uncertain about his eye color, hair consistency, and his complexion. As an artist, you know that your choices are blue, brown, or green eyes; fine, thick, or curly hair; and a fair or ruddy complexion.

How many drawings will you have to complete to include all possible combinations?

Students may figure the possibilities to be 18 (3 x 3 x 2) or may wish to draw the various combinations on paper.

There are many possibilities in the extension area including a lesson on heredity, an art lesson on location of facial features, or social studies lessons on ethnic background and features.

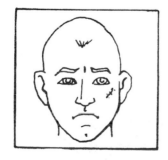 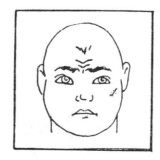 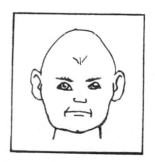

 Hands On, Inc
2121 Rebild Drive
Solvang, CA 93463

34	Uses multiplication to find total number of outcomes

Deli Delight

Grade Level: Upper

MATERIALS: Construction paper; staplers; patterns for lunch meat, bread, and cheese; scissors; and tally sheets

ORGANIZATION: Teams of two, three, or four students

PROCEDURE: Explain to students that they will be building a variety of sandwiches using three types of meats -- salami/turkey/roastbeef, two types of cheese -- Jack/Cheddar, and three types of bread -- rye/wheat/white

Assign a different color construction paper to each item. You may want to get fancy and have students cut out bread, meat, and cheese shaped pieces of paper ahead of time.

Ask students to predict the number of different sandwich combinations that can be created from these ingredients. One rule which students must follow is that each sandwich must have meat, cheese, and bread. Have them create the various combinations and "staple" each sandwich together. There are 18 possible creations.

Upon completion of the project ask students to find a mathematical way of determining the possible combinations. With some guidance, students will realize that:

3 (meat types) x 2 (cheeses) x 3 (breads) = 18

Based upon this information, have students determine the possible combinations with choices of 3/3/3 or 4/3/4, etc.

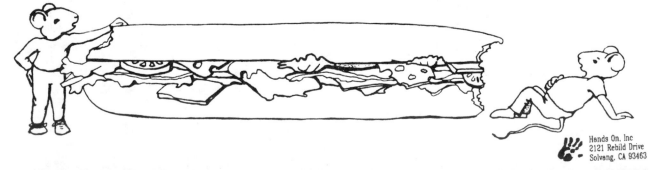

Hands On, Inc
2121 Rebild Drive
Solvang, CA 93463

35	Calculates the probability when choices are dependent or independent

A Real Card

Grade Level: Upper

MATERIALS: Decks of cards divided into groups of four (spades, clubs, hearts, diamonds) so that each group receives thirteen cards – all of the same suit (ace -- king).

ORGANIZATION: Teams of two, three, or four students

PROCEDURE: Before beginning this lesson, explain to students the difference between "dependent" and "independent" events and how they affect probability (explained below).

"Dependent" events are those in which the probability of a selection is influenced by a previous selection.

"Independent" events are those in which the probability of each selection is not reliant on a previous selection.

In this lesson, students will learn about the difference between the two.

Hand out a set of thirteen cards to each group of students. Ask them to identify the probability of selecting a "deuce" out of the stack (the chances are 1 in 13). Have them select one card and place that card aside -- thus creating a stack of 12 cards. What are their chances now of selecting a deuce? (probability is now 1 in 12). The point being that by leaving out a card already selected, the second selection probability is DEPENDENT upon the selection during the first pick.

To emphasize this idea, ask the question, "What are the chances of selecting a deuce on the second pick if the deuce was selected on the first pick?" If the card is left out of the stack, the probability is 0 -- an even greater demonstration of DEPENDENCE.

To demonstrate the INDEPENDENT outcome, have students return the card selected on the first pick and shuffle. The probablity will always be 1 in 13 of selecting any card. In other words, each selection is INDEPENDENT of previous selections.

The purpose of this introductory lesson is to present this unusual concept. The reason for learning about INDEPENDENT and DEPENDENT outcomes is explored in the remaining lessons of this task analysis item.

Hands On, Inc
2121 Rebild Drive
Solvang, CA 93463

35	Calculates the probability when choices are dependent or independent

Just Hanging Around

Grade Level: Upper

MATERIALS: No special materials are needed for this activity

ORGANIZATION: Divide students into pairs

PROCEDURE: This activity is based on the word game "Hangman." Most of your students will know how to play the game, but they will be looking at it from a probability standpoint.

On the board do the following demonstration. Layout the hangman format as pictured. The hidden word will be POWER. Ask students the possibility of selecting one of the five letters used on their first guess (5 in 26). If the guess is right, project that for the next correct letter the probability is 4 in 25; if the guess is wrong the probability is 5 in 25.

You need to point out to students that once letters have been choosen, the possible number of letters for selection gets smaller -- it is therefore a DEPENDENT outcome.

Continue through the game, each time writing the probability of selecting one of the letters used in the word.

Have students play a game between team members. Have them chart the "odds" as they move through the game.

You may also wish to discuss the fact that since letters are sometimes repeated in words, the probabilities will also change based on these repeated letters.

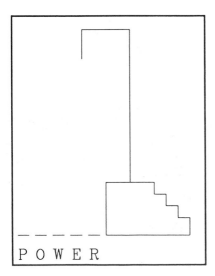

P O W E R

 Hands On, Inc
2121 Rebild Drive
Solvang, CA 93463

35	Calculates the probability when choices are dependent or independent

Losing Your Marbles

Grade Level: Upper

MATERIALS: Four black, four white, and four red marbles and a bag or container for each group

ORGANIZATION: Teams of four

PROCEDURE: Put all the marbles in a container. This is a multi-activity lesson in which students speculate on probability given questions by the teacher.

After asking each question have students predict and record their predictions as a group. Do not allow students to look into the container.

1. If there are twelve marbles in the container, how many marbles must be taken out to be sure that there are at least two marbles of the same color?
2. How many marbles must be removed before you know how many colors of marbles are in the container?
3. If there are three colors in the container, how many marbles must be selected before you get one of each color?
4. If there are three colors in the container, how many marbles must be selected before you get two of each color?
5. How is each of these questions altered if you: a) keep the marble out of the container after you have selected it?; b) place the marble back in the container after you select it?

Have students attempt to create a table, graph, or tree diagram to demonstrate this information.

35	Calculates the probability when choices are dependent or independent

Extra Sensory Perception

Grade Level: Upper

MATERIALS: ESP Cards (described below), tally sheets

ORGANIZATION: Do first as a class lesson then divide students into pairs

PROCEDURE: This is a lesson which tests students for ESP (Extra Sensory Perception) and allows them to predict probable outcomes.

To begin the lesson, show students the ESP cards which they will be using. This deck consists of 25 cards -- five each of cards with a circle, square, star, plus sign, and wavy lines (as in the sample). Tell students to number to twenty-five on a sheet of paper and as you hold each card and concentrate on the symbol, students should write down the "guess" of the symbol on that card. Do this for all twenty-five cards.

Ask students to decide on the probability of being right on any guess (probability is 1 in 5) and ask how many correct responses out of 25 the students might expect to have achieved (5). You might also discuss what ESP is and how this experiment might prove its existence. Correct papers and discuss results.

Divide students into pairs and and have them conduct the experiment with one another.

Have each student complete a paragraph explaining why his partner does or does not have ESP. Students should include information of probability in their responses.

 Hands On, Inc
2121 Rebild Drive
Solvang, CA 93463

36	Reads and makes a scattergram and tells whether it shows positive, negative, or no correlation

Bullseye!

Grade Level: Upper

MATERIALS: Large graph paper (at least one inch squares), beans or markers, large wads of paper (a full sheet) and a target drawn on the chalkboard.

ORGANIZATION: Teams of two, three, or four students

PROCEDURE: Tell students that they are going to make predictions based on a scattergram and ask if there is anyone who has an idea what a scattergram might be. Simply put, a scattergram is a graph with dots of information which records results and forecasts trends.

Each team will need to make two or three paper wads and will need to set up the graph on their desks as shown in the sample. Begin by having students stand one step away from the chalkboard and toss the three paper wads (one at a time) at the target. Record results by placing beans within the corresponding graph square.

Have students then move to a two step distance and complete the same procedure. Continue through a distance of five steps.

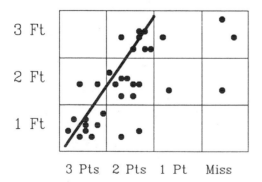

Scattergrams will begin to distribute similar to that shown in the example. Ask students to create a hypothesis based upon the results on their scattergrams -- the greater the distance, the fewer points scored. Show students that there is an "imaginary line" which can be drawn following the concentration of beans, and this line shows trends and can be used to make predictions.

The vocabulary of "positive correlation" can be introduced meaning that the "imaginary line" moves upward and to the right of the graph axis. The negative correlation and no correlation attributes are explained in future lessons.

36	Reads and makes a scattergram and tells whether it shows positive, negative, or no correlation

Making the Grade

Grade Level: Upper

MATERIALS: Graph paper

ORGANIZATION: This is a whole class activity

PROCEDURE: The purpose of this lesson is to introduce students to a scattergram which shows no correlation. The purpose of such a lesson is to allow students to discover that after they have created a hypothesis, data may show that their hypothesis is not true. This is just as valuable as proving a hypothesis.

Ask students to poll one another with the questions: What was your grade point average for your last report card? How much time (on the average) do you spend on homework per night? How many hours of TV do you watch per day? How many times do you wash your hair in a week?

Ask students to create three different hypotheses based upon these four questions. These should include: Students with higher grade point averages watch less TV; students with higher grade point averages spend more time on homework; and students with higher grade point averages wash their hair more often. Based upon the information gathered in the survey, construct three scattergrams on the board. The setup is shown below.

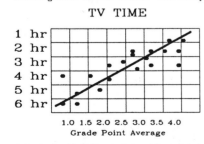

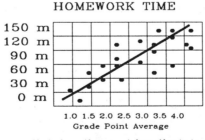

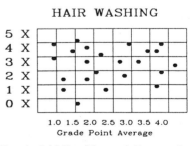

Ask students if the information pictured proves their hypotheses. A hypothesis is proven if a straight line of correlation can be pictured in the graph. An ascending line indicates a positive correlation and a descending line indicates a negative correlation. You will probably find that hypotheses 1 and 2 are generally true while hypothesis 3 shows no correlation.

Have students discuss the value of such an exercise and have them brainstorm other types of hypotheses which will demonstrate no correlation.

The way in which items are labeled on the axis determines the direction of the line. As a general rule your hypothesis dictates the organization of the labels. For example, on the first graph given, the hypothesis is "Students who get better grades watch less television." Therefore, the vertical axis is is labeled from six hours at the base to one hour at the top end; the horizontal axis ranges from a low GPA at the left to a high GPA at the right.

 Hands On, Inc
2121 Rebild Drive
Solvang, CA 93463

36	Reads and makes a scattergram and tells whether it shows positive, negative, or no correlation

What Your Age Will Allow(ance)

Grade Level: Upper

MATERIALS: Pencil, paper, graph paper, a large scattergram for the class

ORGANIZATION: Groups of four

PROCEDURE: Have each group take a different grade level and go to that classroom to interview students as to their age and their allowance.

The hypothesis that students will attempt to prove is that older students receive larger allowances. When students have gathered data for at least five children at each grade level, chart this information on a scattergram on the chalkboard.

Ask each group to decide if there is a positive, negative, or no correlation between age and allowance amount.

An extension might be to interview classes again, but this time ask about bed times for various aged children.

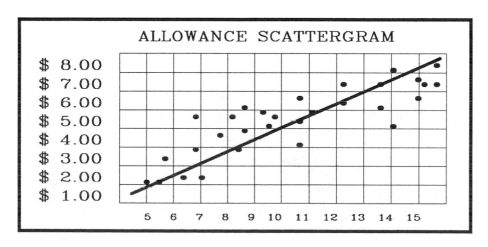

Hands On, Inc
2121 Rebild Drive
Solvang, CA 93463

36	Reads and makes a scattergram and tells whether it shows positive, negative, or no correlation

The Olympics

Grade Level: Upper

MATERIALS: Four or five throwable items of various weights (you might include a softball, baseball, bowling ball, shot put, volleyball, or tennis ball); measuring tape, tally sheets.

ORGANIZATION: A whole class activity to be done in conjunction with pysical education

PROCEDURE: Ask students to speculate on what enables them to throw one item farther than another. Weight or size of an object are obvious factors.

Have each student prepare a tally sheet which lists the five types of items that they will be throwing. Take the entire class out to the playground and begin the exercise by allowing each student to throw each object and measure the distance travelled.

Once the data collection is complete, have students complete two scattergrams. The first should list the objects thrown in order of weight; the second should list the objects in order of size. If you want to include the concept of no correlation, you might have students do a third graph which lists the objects in the order thrown.

Based upon this information, have students create a statement which is proven by the data in their scattergrams.

Distance of Throw By Weight

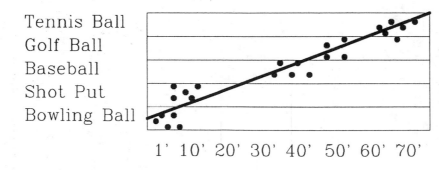

Hands On, Inc
2121 Rebild Drive
Solvang, CA 93463

37	Reads, interprets, and constructs curved graphs

Curved Graph Hunt

Grade Level: Upper

MATERIALS: Newspapers and magazines, especially USA Today, Time, Newsweek, and US News and World Report

ORGANIZATION: Individually or in teams of two

PROCEDURE: Hand out magazines and have students cut out various types of curved graphs. Glue these cutouts on separate sheets of paper and have students create questions about their graphs.

These student generated graph pages can be circulated throughout the room for response and practice by each individual. You may wish to laminate these papers for future use.

Sample Questions:

1. How can you tell which item in the graph is the most popular (most used, most important)?
2. How can you tell which item on the graph is the least popular?
3. How much larger is the largest section than the smallest? How can you figure this out?
4. What is the value of the most popular (most used, most important) item on the graph?
5. Which item on the graph is used an "average" amount of time? How do you know?
6. What are the benefits of putting this information in graph form?
7. Do you find graphs to be a helpful way of presenting information? Why?

There are several task analysis items involving "reading and interpreting" various types of graphs. The intent of this book is to provide activity based math lessons and in general it is difficult to do reading and interpreting in an activity approach; therefore, we have provided universal lessons for these T.A. items and suggest you use these lessons in conjunction with your basic math series or with duplicated material.

Hands On, Inc
2121 Rebild Drive
Solvang, CA 93463

Weather Graphs

Grade Level: Upper

MATERIALS: Copies of five days of weather reports from a newspaper, graph paper

ORGANIZATION: Pairs of students

PROCEDURE: Hand out the information on weather for five days. Ask students to analyze the information and as a team try to decide on information which could be presented as a curved graph.

As a class, discuss each team's selection as to why it will or will not work as a graph. Encourage students to try different approaches with different types of data. There are a variety of acceptable approaches.

Ask each team to plan its graph in terms of the value of the horizontal or vertical axis. Have each team draw its plan on the board. Discuss these plans with the entire class and then complete two or three of the graphs on the chalkboard or overhead.

From Table to Graph

Grade Level: Upper

MATERIALS: Table presented in appendix C, graph paper

ORGANIZATION: Can be done individually or in groups of two, three, or four

PROCEDURE: Hand out copies of appendix C to each student or group. Tell them that they will be constructing a curved graph which depicts this information.

Essential elements to explain include: a title for the graph, labeling of various sectors, neatness, and accurate transfer of data.

You may wish to work with the students in a step by step procedure or simply present the concept and have each group work together to arrive upon a finished product.

Hands On, Inc
2121 Rebild Drive
Solvang, CA 93463

37	Reads, interprets, and constructs curved graphs

Watching TV Watchers

Grade Level: Upper

MATERIALS: Graph paper, paper, and pencils

ORGANIZATION: Can be done individually or in pairs of students

PROCEDURE: In teaching a curved graph it is helpful if you explain to students that curved graphs generally show information over a span of time and that changes are smooth rather than erratic or comparative.

Use some examples such as graphing the amount of sunlight over a twenty-four hour period or the cycle of the moon over a twenty-eight day period.

Once students understand the concept, tell them that they will be constructing a curved graph identifying TV viewing time of the members of their family over a one day period.

Explain that generally all members of the family do not sit down at the beginning of a particular show -- there is an addition of one person at a time.

Have each student complete a tally sheet similar to the model shown. Give them a day or two to complete the information gathering.

When students return with the information have each student construct a curved graph which depicts the viewing habits of his family. You may wish to display this information or transfer the finished graphs to overhead transparencies and overlay the graphs to make comparisons. Discussion of range, median, and mode will fit well into this discussion.

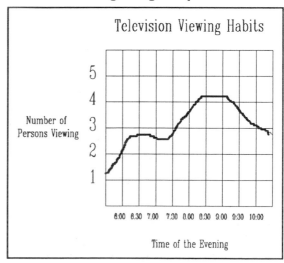

| 37 | Reads, interprets, and constructs curved graphs |

Visitation Rights

Grade Level: Upper

MATERIALS: Pencil and paper

ORGANIZATION: Individually or in groups of two

PROCEDURE: Students will be involved in an "observation" of a certain area of the school. This might be the lockers, the bathrooms, the lunch lines, the baseball field, the apparatus, etc.

Over the course of one day, students should view their area for three to five minutes before school, at recess, lunch, after lunch, and after school. While watching their territory, students should record the number of students using the area during the three to five minute spans.

After information has been gathered, students should create a curved graph by plotting the points of use on the "Y" axis and the time of day on the "X" axis.

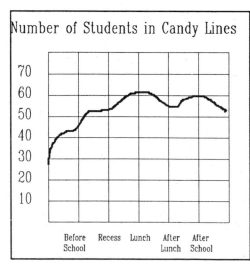

Have students explain the results of their campus "traffic pattern" and suggest possible explanations for their results.

An important point to cover is that a curved graph is apropos in this situation because the traffic patterns ease slowly from one number to the next -- students don't instantly disappear at a given point in time.

Hands On, Inc
2121 Rebild Drive
Solvang, CA 93463

38	Identifies and constructs appropriate graphs for given information

Happy Holidays

Grade Level: Upper

MATERIALS: Large art paper (18x24), colored pens, rulers, yardsticks, and compasses

ORGANIZATION: Groups of four

PROCEDURE: This is a four day project to use prior to any holiday season. The following lessons are focused on the Christmas season.

ACTIVITY 1 – DATA COLLECTION
Ask students to write a list of their:
1. Five favorite Christmas songs
2. Five presents they want to receive
3. Five items that appear on their holiday dinner
4. Five activities that are associated with Christmas (buying, wrapping, singing, etc.)

In their groups of four, generate a list of 5 to 10 items most frequently mentioned in each category.

Representatives from each group meet together to form a final list of 5 to 10 items for each category. This group then makes ballots (as shown) on which students can write their names to vote for their favorites.

☐ Deck the Halls	☐ Oh Come All Ye Faithful
☐ Jingle Bells	☐ Frosty the Snowman
☐ Silent Night	☐ Twelve Days of Christmas
☐ Joy to the World	☐ The First Noel
☐ We Three Kings	☐ Rudolph the Red Nosed Reindeer
☐ Silver Bells	☐ Away in a Manger
☐ Feliz Navidad	☐ Oh Little Town of Bethlehem

Hands On, Inc
2121 Rebild Drive
Solvang, CA 93463

ACTIVITY 2 – GRAPH CREATION

Using the ballots, each group will put data into graphs. Each group can decide on the type of graph which best displays the given information -- circle graph, bar graph, line graph, pictograph, etc. One of each kind should be made by each group. Let students know that their graphs will be evaluated by classmates so they need to be done neatly and large enough for public display.

ACTIVITY 3 – EVALUATION

Have students look at all displayed graphs. As a class, decide upon the things that should be on a ,well-made graph. These should include: accuracy, neatness, comprehensibility, spacing, interest, originality, labels, color, title, and organization. Once the class has reached consensus on ten items for evaluation, have each team evaluate and award a maximum of ten points for each item on the "well-made graph" list. Total possible points will be 100 for each graph on display. Students should have a scoresheet to record their scoring.

ACTIVITY 4 – AVERAGING AND GRADING

On large charts posted on the board (see sample), each group should record the number of points given for each "well-made graph" attribute.

From this composite of scores, each group will calculate its grade by finding the mean (average) for each attribute and then adding these ten averages together. This information should be written on a scoresheet and given to the teacher for grading purposes.

GRAPH SCORING SHEET

☐ Accuracy

☐ Neatness

☐ Comprehensibility

☐ Spacing

☐ Interest

☐ Originality

☐ Labels

☐ Color

☐ Title

☐ Organization

Hands On, Inc
2121 Rebild Drive
Solvang, CA 93463

38	Identifies and constructs appropriate graphs for given information

Take Your Choice

Grade Level: Upper

MATERIALS: Atlases, almanacs, encyclopedias, or other information on states dealing with population, square miles, population of capital cities, etc.

ORGANIZATION: This activity can be done individually or in groups of two, three, or four

PROCEDURE: Discuss the various forms of graphs -- bar, circle, broken line, pictograph. Try to elicit the statement that different graphs serve different functions.

Part of their assignment today will be deciding upon which type of graph would most effectively depict information containing large numbers.

Have each student or group select seven states to research. They might decide to do the population of each state, the number of square miles, the population of the capital city or the largest city, or other information which will allow comparison of information involving large numbers.

Once all data has been collected, have them decide the best way to display this information. Each group should complete a graph in a format which can be posted and discussed.

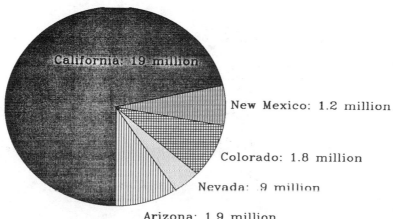

38	Identifies and constructs appropriate graphs for given information

Billboard Bonanza

Grade Level: Upper

MATERIALS: Top 40 listing from Billboard Magazine for the past several weeks, graph paper

ORGANIZATION: Individually or in groups of two, three, or four students

PROCEDURE: This is a multi-faceted lesson with a tremendous amount of high interest information for graphing. Using the information from one or several of the Billboard pages, have students create three different types of graphs -- bar, line, broken line, circle, pictograph, or curved graph.

Most students will not have seen a copy of Billboard Magazine, and they will be fascinated with the various articles and charts. Almost any activity described in the task analysis for statistics, probability and graphing has possible lessons using Billboard.

As an extension, you might use Billboard for writing news stories, graphic layout as related to art, and a variety of other activities.

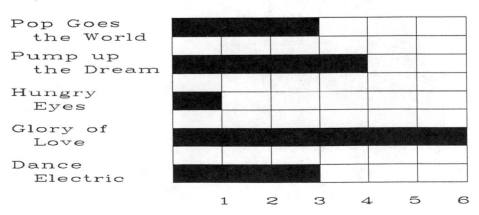

Number of Weeks on the Chart

 Hands On, Inc
2121 Rebild Drive
Solvang, CA 93463

Materials Lists

Primary Grades Materials	Middle Grades Materials	Upper Grades Materials
Apples	Almanacs	Almanacs
Bowls	Atlases	Atlases
Butcher Paper	Bathroom Scale	Balls – an assortment
Butter	Beans	Bathroom Scale
Candy (wrapped, hard)	Blocks of Wood	Beaker – small
Candy Conversation Hearts	Bottle Caps	Beans (bags of uncooked)
Construction Paper	Butcher Paper	Bowls
Counting Cups – Portion Cups	Cassettes	Bucket (large beaker)
Crayons	Chips (several colors)	Candy (Lifesavers)
Egg Cartons	Colored Pencils	Cards (several decks)
Glue	Construction Paper	Catalogues for clothing/gifts
Graphing Plastic	Data Watches	Chips (Poker)
Grid Sheets (1″ – Appendix E)	Dice	Coins
Hard–Boiled Eggs	Glue	Compasses
Kites (3)	Graph Paper	Construction Paper
Knife	Hammers	Crayons
Lima Beans	Magazines	Cylinders (able to roll)
Marking Pens	Masking Tape	Dice (regular and polyhedral)
Marshmallows	Marking Pens	Encyclopedias
Masking Tape	Nails (6D, 8D, 10D)	ESP Cards
Milk Cartons (school)	Newspapers	Eyedroppers
Paint Brushes	Play Money (nickels)	Financial page of newpapers
Peanut Butter	Paper Cups	Glue
Peanuts (can)	Pencils	Graph Paper (various sizes)
Peanuts (in shell)	Perpetual Calendars	Jars (assorted)
Pencils	Recording Sheets	Magazines (Time/Newsweek/etc
Picture of Abe Lincoln	Rulers	Magazines (Billboard)
Plastic Cups	Scissors	Magnets
Popcorn and popper	Spinners (Appendix A)	Markers
Potting Soil	Stop Watches	Marbles
Pumpkins	String	Masking Tape
Salt	Stumps of Wood	Measuring Tape
Scissors	Tally Sheets	Measuring Cups
Starch	Telephone Books	Newspapers
String	Temperature Charts	Notecards (3x5)
Styrofoam cups		Play Money
Tag Board		Rulers
Template (1–10)		Scissors
Tissue Paper		String
Tongue Depressors		Stopwatch
Toothpicks		Washers
Unifix Cubes		Yard or metric sticks

Graphing Questions

PRIMARY

1. What number sentence can you tell me about the graph?

2. Which column has the least?

3. Which column has the most?

4. Are there less ＿＿＿ or less ＿＿＿ ?

5. Are there more ＿＿＿ or more ＿＿＿ ?

6. How many ＿＿＿ are there?

7. How many less ＿＿＿ are there than ＿＿＿ ?

8. How many more ＿＿＿ are there than ＿＿＿ ?

9. How many ＿＿＿ are there all together?

10. Are any columns the same?

Making a Spinner

Many activities in this book require the use of spinners. Students can make their own spinners very easily and inexpensively by following the directions given here.

Materials required are: cardboard, scissors, rulers, pencils, paper clips, paper punch, and tape.

1. Cut out a cardboard (tagboard) pointer and punch a hole in the center.

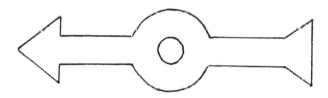

2. Cut a scrap of cardboard as a paper washer and punch a hole in the center.

3. Cut out a four inch square of cardboard and divide it into quarters as shown. Make light pencil lines.

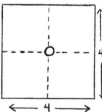

4. At the center of the square, make a small hole with your paper clip.

5. Using a compass, make a circle on the four inch square and draw the design you wish to use. You may wish to color your spinner at this point.

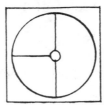

6. Bend the center loop of the paper clip up at a 90 degree angle to the outer loop.

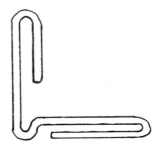

7. Tape the outer loop of the paper clip to the bottom of the four inch square to hold it in place.

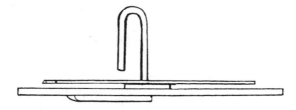

UNDERSIDE OF SPINNER

8. Put the bent paper clip through the hole in the four inch square, the paper washer and the pointer.

THE FICTITIOUS FAMILY FILE

The fictitious family file is an integrated activity used in Task Analysis item 33 (Construct a circle graph given data). It lends itself to a one to two week unit involving predicting, budgeting, creating circle graphs, and comparing expenditures.

There are thirty-three sample families described on the back of this page. Each student should draw one of these families from a box. Some families will be single parents, some are large, some small. Place all of the "single family male households" and half of the other families in one box for the boys to draw from. Place the "single family female households" and the other half of the families in another box for the girls to draw from.

Each family has a stated yearly income based upon the job listed.

Each student should use the information about his family to complete the tasks, which include: income expenditures (budget), holiday expenditures (based on savings), vacation expenditures (based on savings), and summer clothing (based on clothing allocation).

This unit can be done as a basic plan using only the materials provided in the lesson plans; or, it can be extended to include a variety of catalog searches, guest speakers, trips to a local travel agent or real estate office, or extensions into various other curricular areas.

Another extra activity which you may choose to use is to create a collage or family album by having the students cut pictures from a magazine to graphically illustrate his imaginary family.

Model Family Cards

The GREEN Family Father, Mother, Girl 7, Boy 15 Father: Dentist, $54,000 Mother: None	**The SMITH Family** Father, Boy 4, Girl 12 Father: Architect, $46,500	**The RODGERS Family** Father, Mother, Boy 2 Father: Attorney, $65,000 Mother: Designer, $65,000
The ADAMS Family Father, Mother, Boy 9, Girl 15 Father: Plumber, $36,000 Mother: None	**Ms. HIGHSMITH** Single Woman Librarian: $19,700	**The EVANS Family** Father, Mother, Girl 3, Girl 5 Father: Draftsman, $23,000 Mother: Owns Business, $30,000
The ALLAN Family Father, Mother, Girl 15 Father: Electrician $31,000 Mother: Store Clerk, $8,500	**The JOHNSON Family** Father, Mother, Boy 7, Boy 13 Father: Photographer $19,600 Mother: None	**The WILLIAMS Family** Father, Mother, Girl 1, Boy 7, Girl 15 Father: Mechanic, $17,900 Mother: None
The CRAWFORD Family Father, Mother, Boy 8, Girl 10, Boy 13 Father: Salesman, $18,500 Mother: None	**The ALLSMILLER Family** Mother, Girl 16 Mother: Nurse $30,300	**The ROARK Family** Husband, Wife Husband: Fireman, $35,000 Wife: Teacher, $22,000
Mr. SHELBY Single Man Carpenter, $25,000	**The DRAKE Family** Father, Mother, Boy 7, Boy 15 Father: Painter, $26,000 Mother: Secretary, $15,400	**Mr. SHELBY** Single Man Doctor, $75,000
The IRWIN Family Father, Mother, Boy 1, Girl 7 Father: Mailman, $21,000 Mother: Store Clerk, $9,700	**The KETCHUM Family** Father, Boy 9, Boy 16 Father owns tire store $37,900	**The RAY Family** Mother, Girl 9 Owns Dress Shop, $29,300
The COLLINS Family Father, Mother, Boy 7, Boy 9, Girl 17 Father: Plant Manager, $46,000 Mother: Nurse, $26,000	**Ms. HERNANDEZ** Woman Insurance Salesperson, $24,000	**The WATSON Family** Father, Mother, Girl 7, Boy 11 Father: Truck Driver, $21,300 Mother: Secretary, $10,800
The CROWE Family Father, Mother, Boy 15, Girl 18 Father: Doctor, $81,000 Mother: None	**The REDFORD Family** Mother, Girl 7, Girl 11, Girl 15 Real Estate Agent, $32,000	**The SHARK Family** Father, Mother, Boy 5 Father: Teacher, $29,000 Mother: Teacher, $31,000
The O'MARA Family Father, Mother, Boy 7, Girl 9, Boy 13 Father: Artist, $23,000 Mother: Lawyer, $35,000	**The REAGAN Family** Father, Mother, Girl 6, Boy 12 Father: Fireman, $25,600 Mother: None	**The OLIVIA Family** Father, Mother, Boy 8, Girl 12 Father: Bank VP, $45,000 Mother: None
The ORTEGA Family Father, Mother, Girl 6, Girl, 8, Boy 12 Father: Policeman, $28,000 Mother: Secretary, $13,500	**The SHELL Family** Father, Boy 4, Boy 10 Father: Bank Teller, $18,500	**The BENNETT Family** Mother, Girl 6, Girl 14 Mother: Waitress, $10,200
The O'GRADY Family Father, Mother, Boy 2, Girl 4, Girl 8, Boy 12 Father: Manager, $24,000 Mother: None	**The NELSON Family** Father, Boy 4, Girl 11 Father: Teacher, $31,000	**The OBRIEN Family** Father, Mother, Boy 12, Boy 17, Boy 19 Father: Store Owner, $65,000 Mother: None

TEMPERATURE COMPARISON GRAPH

| MONTH | NEW YORK TEMPERATURE | | | CHICAGO TEMPERATURE | | | LOS ANGELES TEMPERATURE | | |
	MAX	MIN	PRECIP.	MAX	MIN	PRECIP.	MAX	MIN	PRECIP.
JANUARY	32	19	19	33	17	10	65	45	7
FEBRUARY	32	17	17	35	20	10	66	47	5
MARCH	41	25	16	45	29	12	69	49	6
APRIL	53	34	14	58	39	13	71	52	4
MAY	66	45	13	70	49	12	74	55	2
JUNE	76	56	11	80	59	10	77	58	1
JULY	81	60	10	85	64	9	83	62	0
AUGUST	79	59	10	83	62	8	84	62	1
SEPTEMBER	73	52	11	76	55	8	82	60	1
OCTOBER	60	42	12	64	44	7	77	56	2
NOVEMBER	47	33	15	48	31	10	73	51	4
DECEMBER	35	23	18	35	21	10	67	48	6

Glossary of Terms

Array	An orderly arrangement of objects or symbols in rows and columns.
Average	The sum of all the items in a set of data divided by the number of items. Synonymous with **mean**.
Bar Graph	A graph in which the height or length of each bar is proportional to the size of the data item it represents.
Broken Line Graph	A graph of lines connecting marked points on a grid of horizontal and vertical lines showing a change over a period of time.
Census	An official count of all the people in a given population.
Circle Graph	A circle used to show how a whole quantity is divided into parts. Usually displaying data in fraction, decimal, or percent form.
Collate	Arrange in order or put together in specific order.
Complementary Events	Two mutually exclusive events, one of which must occur to allow the other to occur. The sum of the probabilities of complementary events is 1.
Counting Cups	1/2" high by 2" diameter paper cups sold in supply stores.
Curved Graph	Similar to a **broken line graph** but the change between events occurs gradually and is therefore depicted by curved lines rather than straight lines. A bell–curve is an example of a curved graph.
Data	Facts about objects or events. Synonymous with **information**.
Dependent Events	Events in which the outcome of first choice affects the probabilities of the outcome of the second event (e.g. The chance of selecting the ace of spades from a full deck of cards is 1 in 52. If one card has already been selected and not replaced, the chances are 1 in 51).
Diagram	A drawing or sketch that shows the relationship of its parts.
Double Bar Graph	A bar graph in which parallel bars compare the differences between data (e.g. population in 1987 vs. population in 1988).
Estimate	To judge an approximate worth, size, or amount.

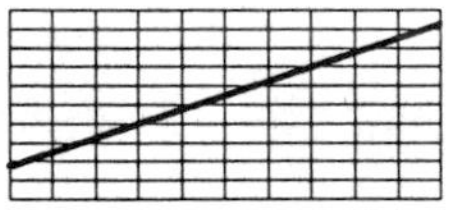

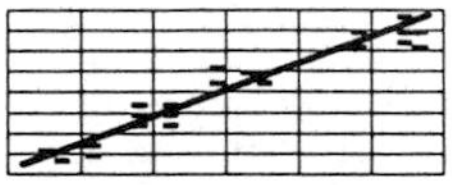

Exclusive Event	An event which is limited to itself and has no bearing on other events.
Frequency Table	A table that organizes numerical data into intervals that can be tallied to show frequency and relative frequency.
Generate	To gather information.
Graph	A drawing designed to show relationships between two (or more) sets of values.
Graph Key	Information that makes a graph understandable.
Graphing Plastic	60" by 90" plastic with squares marked off with masking tape in 3 or 4 columns. Can be made of vinyl or plastic garbage bags.
Histogram	A bar graph that shows the frequencies of intervals of data.
Horizontal Axis	A line used for reference going left to right. Also known as the "X-axis."
Hypothesis	An assumption based on established fact.
Interpret	To explain the meaning of.
Intersection	A point, line, or place where one item crosses another.
Legend	The explanation of the symbols or numerals used on a graph.
Line Graph	Guide line plotted through points of a graph. A line graph shows a trend.
Line of Best Fit	A straight line on a scattergram where most points lie.
Mean	The sum of the numbers divided by the number of addends. Synonymous with **average**.
Median	The MIDDLE number when a set of numbers are arranged in order. It may be, but is not necessarily, the **average**.
Mode	The number occurring most often in a set of data.
Negative Correlation	A statistical term in which one variable increases as the other variable decreases.

No Correlation	A statistical term in which there is no relationship between the variables.; neither positive nor negative correlation.
One to One Correspondence	A pairing which matches one member of a set with one and only one member of another set.
Original Data	Information created for a graph or a table.
Pattern	A regular arrangement in a sequence or order.
Permutations	Any changes in "order" possible within a set. A tree diagram shows possible permutations.
Perpetual Calendar	A never ceasing, continuous calendar.
Pictograph	A diagram, chart, or graph presenting statistical data by using pictures, symbols, or different colors, sizes, or numbers.
Polyhedral Dice	Dice having more sides than standard dice.
Population	In statistics, the entire group of items or individuals from which the samples under consideration are presumed to come.
Positive Correlation	A statistical term in which both variables increase or both decrease.
Possible Outcomes	All results that may exist.
Probability	Something likely to happen -- the likelihood that an event will occur estimated as a ratio as, in "1 of 2" or "1 out of 2."
Proportion	The relation in size, number, or amount, of one thing compared to another.
Random	By chance -- without a definite aim, plan, method, or purpose.
Range	The difference between the largest and smallest number in a set of data.
Ratio	A comparison of two members by division -- 3/2, 3:2, 3÷2.
Raw Data	A collection of facts that have not been processed into usable information.

Scattergram | A graph of ordered pairs of points showing positive, negative, or no correlation between two sets of data.

Sample | Data gathered from a part of a population which is representative of the response of the entire population.

Sector | A defined area -- a section or portion of a graph.

Statistics | The science of collecting, classifying, and analyzing facts.

Symbol | A letter, figure, or sign that stands for or represents an idea, quality or object.

Table | Information in a brief, tabular form.

Tabular Form | Data arranged for quick reference as in columns.

Tally | A mark made to record a certain number of objects -- used for keeping count.

Template (1 - 10) | Heavy tagboard with a hole below each number -- cut out with manicure scissors.

Tree Diagram | A diagram used to find the total number of possible outcomes in a probability experiment.

Trend | The direction, course, or tendency shown by the data.

Unifix Cubes | Smooth plastic cubes which snap together firmly.

Vertical Axis | A line used for reference going from top to bottom -- also known as the Y-axis.

Zero Probability | An event which cannot happen.